Lecture Notes in Computer Science 14328

Founding Editors

Gerhard Goos
Juris Hartmanis

The series Lecture Notes in Computer Science (LNCS), including its subseries Lecture Notes in Artificial Intelligence (LNAI) and Lecture Notes in Bioinformatics (LNBI), has established itself as a medium for the publication of new developments in computer science and information technology research, teaching, and education.

LNCS enjoys close cooperation with the computer science R & D community, the series counts many renowned academics among its volume editors and paper authors, and collaborates with prestigious societies. Its mission is to serve this international community by providing an invaluable service, mainly focused on the publication of conference and workshop proceedings and postproceedings. LNCS commenced publication in 1973.

Muge Karaman · Remika Mito ·
Elizabeth Powell · Francois Rheault ·
Stefan Winzeck

Editors

Computational Diffusion MRI

14th International Workshop, CDMRI 2023
Held in Conjunction with MICCAI 2023
Vancouver, BC, Canada, October 8, 2023
Proceedings

 Springer

Editors
Muge Karaman 🆔
University of Illinois at Chicago
Chicago, IL, USA

Elizabeth Powell 🆔
University College London
London, UK

Stefan Winzeck 🆔
Microsoft Research Cambridge
Milton, UK

Remika Mito 🆔
Florey Institute of Neuroscience and Mental
Health
Heidelberg, VIC, Australia

Francois Rheault 🆔
Université de Sherbrooke
Sherbrooke, QC, Canada

ISSN 0302-9743   ISSN 1611-3349 (electronic)
Lecture Notes in Computer Science
ISBN 978-3-031-47291-6   ISBN 978-3-031-47292-3 (eBook)
https://doi.org/10.1007/978-3-031-47292-3

This Springer imprint is published by the registered company Springer Nature Switzerland AG
The registered company address is: Gewerbestrasse 11, 6330 Cham, Switzerland

Paper in this product is recyclable.

Preface

We are delighted to introduce the proceedings of the 2023 Computational Diffusion MRI (CDMRI) Workshop. CDMRI has been running as a satellite workshop of the Medical Image Computing and Computer Assisted Interventions (MICCAI) conference for over ten years. This platform has consistently provided the opportunity for researchers from around the globe to present and discuss the latest developments in the acquisition, analysis, and application of diffusion MRI.

Over the past four decades, major strides in diffusion MRI (dMRI) research have transformed our ability to non-invasively probe tissue microstructure. The dMRI research community has contributed major advances spanning from novel acquisition and modeling methods to clinical insights from the application of quantitative dMRI methods. In recent years, machine learning techniques have begun to help tackle the challenges of acquiring rich datasets in clinically feasible times, as well as to make sense of big data to answer clinical questions. The submissions to CDMRI 2023 showcase this diverse range of research within the field of dMRI. These contributions cover various aspects, including preprocessing, signal modeling, tractography, bundle segmentation, and clinical applications. Many of these studies employ novel machine learning implementations, highlighting the evolving landscape of techniques beyond the more traditional physics-based algorithms.

The now immense body of research in dMRI has showcased the clinical potential of this technique; however, one of the major challenges for clinical implementation is in the inherent variability in dMRI measures across sites and scanners. Differences in acquisition, scanner hardware, and site result in variability in dMRI measures, making clinical interpretation difficult. Consequently, a major current focus in the dMRI community is on harmonization of dMRI data across sites and scanners, as without appropriate harmonization techniques, it will be difficult not only to scale up dMRI from the single site to the broader community, but additionally to make use of big multi-site datasets that are becoming increasingly available. This year, the CDMRI Workshop hosted the MICCAI Challenge entitled "Quantitative Connectivity through Harmonized Preprocessing of Diffusion MRI" (QuantConn). The QuantConn organizers tasked participants with making data from the two acquisitions of a paired dataset (n = 100) as similar as possible, using any means they wished (e.g., denoising approaches, super resolution, explicit image harmonization approaches), with the goal of retaining biological differences while mitigating any acquisition-related differences.

This workshop would not have been possible without the dedication of the Program Committee (listed below), who, through a double-blind peer review process, ensured the highest standard of submissions was achieved. Of the 19 submissions we received, 9 were accepted with minor revision, 8 were accepted after revisions, and 2 were rejected; each submission was reviewed by at least two members of the Program Committee. We extend our gratitude to everyone involved in the Program Committee. Finally, we wish to thank our Keynote Speakers (listed below), who delivered a series of truly enlightening lectures.

September 2023

Muge Karaman
Remika Mito
Elizabeth Powell
Francois Rheault
Stefan Winzeck

Organization

CDMRI Organisers

Muge Karaman — University of Illinois at Chicago, USA
Remika Mito — Florey Institute of Neuroscience and Mental Health, Australia
Elizabeth Powell — University College London, UK
Francois Rheault — Université de Sherbrooke, Canada
Stefan Winzeck — Microsoft Research Cambridge, UK

CDMRI Program Committee

Nagesh Adluru — University of Wisconsin-Madison, USA
Suyash Awate — Indian Institute of Technology Bombay, India
Leon Cai — Vanderbilt University, USA
Maxime Chamberland — Eindhoven University of Technology, The Netherlands
Bramsh Chandio — University of Southern California, USA
Daan Christiaens — KU Leuven, Belgium
Marta Correia — MRC Cognition and Brain Sciences Unit, UK
Samuel Deslauriers-Gauthier — Inria Centre at Université Côte d'Azur, France
Matteo Figini — University College London, UK
Sila Genc — Royal Children's Hospital, Australia
Jon Haitz Legarreta Gorroño — Brigham and Women's Hospital, USA
Jana Hutter — King's College London, UK
Sila Kurugol — Boston Children's Hospital and Harvard Medical School, USA
Christophe Lenglet — University of Minnesota, USA
Jinglei Lv — University of Sydney, Australia
Lipeng Ning — Harvard Medical School, USA
Marco Palombo — Cardiff University, UK
Tomasz Pieciak — Universidad de Valladolid, Spain
Marco Pizzolato — DTU, Denmark
Simona Schiavi — ASG Superconductors S.p.A., Italy
Thomas Schultz — University of Bonn, Germany
Etienne St-Onge — Université du Québec en Outaouais, Canada
Antoine Théberge — Université de Sherbrooke, Canada

Ye Wu	Nanjing University of Science and Technology, China
Joseph Yuan-Mou Yang	Royal Children's Hospital, Australia
Qianqian Yang	Queensland University of Technology, Australia
Pew-Thian Yap	UNC Chapel Hill, USA
Zheng Zhong	Stanford University, USA

Keynote Speakers

Jennifer A. McNab	Stanford University, USA
Demian Wassermann	Inria, Université Paris-Saclay, France
Chun-Hung Yeh	Chang Gung University, Taiwan

Contents

Neural Spherical Harmonics for Structurally Coherent Continuous Representation of Diffusion MRI Signal

Tom Hendriks$^{(\boxtimes)}$ (ID), Anna Vilanova (ID), and Maxime Chamberland (ID)

Department of Computer Science and Mathematics, Eindhoven University
of Technology, Groene Loper 5, 5612 AP Eindhoven, The Netherlands
t.hendriks@tue.nl

Abstract. We present a novel way to model diffusion magnetic resonance imaging (dMRI) datasets, that benefits from the structural coherence of the human brain while only using data from a single subject. Current methods model the dMRI signal in individual voxels, disregarding the intervoxel coherence that is present. We use a neural network to parameterize a spherical harmonics series (NeSH) to represent the dMRI signal of a single subject from the Human Connectome Project dataset, continuous in both the angular and spatial domain. The reconstructed dMRI signal using this method shows a more structurally coherent representation of the data. Noise in gradient images is removed and the fiber orientation distribution functions show a smooth change in direction along a fiber tract. We showcase how the reconstruction can be used to calculate mean diffusivity, fractional anisotropy, and total apparent fiber density. These results can be achieved with a single model architecture, tuning only one hyperparameter. In this paper we also demonstrate how upsampling in both the angular and spatial domain yields reconstructions that are on par or better than existing methods.

Keywords: Diffusion MRI · Implicit Neural Representation · Spherical Harmonics

1 Introduction

The human brain is a highly structured organ. With the introduction of diffusion magnetic resonance imaging (dMRI) in vivo study of the structure of the brain became a possibility. The spatially coherent structures in the brain imply that spatial coherence should be present when modeling dMRI data. Diffusion tensor imaging (DTI) [3] fits a tensor for every voxel of the volume describing the diffusion in three primary directions. Constrained spherical deconvolution (CSD) [14] can describe the orientation and relative size of fiber bundles using fiber orientation distribution functions (fODFs). These are examples of methods that model the fiber orientation in every voxel independently, disregarding any

intervoxel coherence. Interpolating correctly between voxels using classical interpolation methods (e.g. cubic interpolation) is, therefore, difficult and susceptible
to noise, and can discard anatomical details. Interpolation in the angular domain
has proven to be a difficult task as well, as highlighted by recent challenges in
the computational dMRI community [4,9,13]. Machine learning approaches for
upsampling in both the angular and spatial domains are a promising avenue [1].
However, these methods often rely on a strong prior obtained by training on
large amount of data. This is problematic when training data is scarce or if the
model is applied to data inherently different from the data it was trained on
(e.g. pathological data). Ideally, a continuous and structurally coherent model
should be derived at the individual level (i.e., n = 1).

Neural radiance field (NeRF) [8] models have shown to be extremely effective
at creating continuous 3-dimensional representations, known as implicit neural
representations, of scenes given a limited number of 2-dimensional input images
taken from limited angles. NeRF overfits a multi-layer perceptron to essentially
capture a given scene in its parameters. Unseen angles can then be sampled from
this network. This concept could translate well to dMRI, as we are trying to
create a complete representation from an incompletely sampled angular domain.
The difference with dMRI is that every angle in a dMRI-acquisition produces a
complete 3-dimensional volume of data.

In this work, we propose to use a NeRF-like model to create a model of
the dMRI data of a single subject that utilizes the structural coherence of the
brain, while providing continuity in both the angular and spatial domain. We
evaluate the resulting model in a number of downstream tasks, such as calculating microstructural metrics, and fODF estimation. We also demonstrate how
the model can be used to upsample dMRI data in both the angular and spatial
domain.

2 Methods and Experiments

2.1 Data

We sourced data from a single participant from the preprocessed Human Connectome Project dataset [18] consisting of 18 $b = 0$ s/mm^2 volumes, 90 $b = 1000$
s/mm^2 volumes, 90 $b = 3000$ s/mm^2 volumes, with 1.25 mm isotropic voxels.

2.2 Model

The neural spherical harmonics model (NeSH) is an adaptation from SH-
NeRF [19]. NeSH outputs an approximation $\hat{S}(x, y, z, \boldsymbol{b})$ for a diffusion signal
$S(x, y, z, \boldsymbol{b})$. An input pair $i \in I$ consists of a voxel midpoint coordinate (x, y, z)
and a gradient direction vectors $\boldsymbol{b}$, where I is a set of all possible coordinate-
direction pairs. I has size $N = n_c \times n_d$ with n_c being the number of coordinates
and n_d the number of directions. The input coordinates are scaled to lie in

$[-1,1]$ and are positionally encoded using the generalization of the NeRF positional encoding [12] into input vector $\boldsymbol{x}$. Direction vector $\boldsymbol{b}$ is converted into the corresponding polar angles θ (azimuth, $[0, 2\pi)$) and ϕ (elevation, $[0, \pi]$).

A simple multi-layer perceptron (MLP) maps $\boldsymbol{x}$ into a coefficient vector $\boldsymbol{k}$ that parameterizes a spherical harmonics (SH) series.

$$M_\Psi : \boldsymbol{x} \rightarrow \boldsymbol{k} \tag{1}$$

$$\boldsymbol{k} = (k_l^m)_{l:0 \leq l \leq l_{max}}^{m:-l \leq m \leq l} \tag{2}$$

where k_l^m is the coefficient for the SH component of degree l and order m, l_{max} is the maximum degree of the SH-series, and m the order. For a given i we can now obtain an estimation of the dMRI signal:

$$\hat{S}_i = \hat{S}(x_i, y_i, z_i, \boldsymbol{b}_i) = \sum_{(k_l^m)^i \in \boldsymbol{k}_i} (k_l^m)^i Y_l^m(\theta, \phi) \tag{3}$$

where $Y_l^m(\theta, \phi)$ is the SH component of degree l and order m for azimuth θ and elevation ϕ obtained from $\boldsymbol{b}_i$, $\boldsymbol{k}_i$ is the coefficient vector given by (1) for input i, and $(k_l^m)^i$ is the coefficient of Y_l^m. The dMRI signal is reconstructed with a simplified real basis SH series, using only odd-numbered degrees. Different methods of simplifying the SH series exist [5,14]; in this paper, the method described in MRtrix3 is used [17]. The full model is shown in Fig. 1.

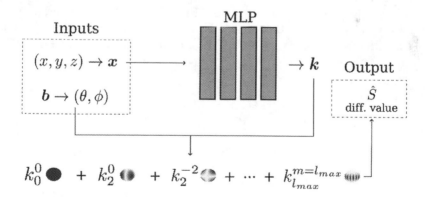

Fig. 1. A schematic representation of the Neural spherical harmonics (NeSH) model. Inputs coordinates are spatially encoded into (x), directional vector $\boldsymbol{b}$ is converted to θ and ϕ. Vector $\boldsymbol{x}$ is passed through the multi-layer perceptron (MLP) to produce $\boldsymbol{k}$, which parameterizes the spherical harmonics series. This is sampled in direction $\boldsymbol{b}$ to produce the final output $\hat{S}$.

We calculate the loss as an average over all inputs for the smooth L1 loss [6] between the value of $\hat{S}_i$, and the dMRI signal S_i defined as the dMRI signal measured at (x_i, y_i, z_i) in the direction of $\boldsymbol{b}_i$. Unregularized, NeSH could be susceptible to overfitting on noise, if the maximum degree of the SH series is larger than necessary to model the diffusion data in a given voxel. An L1 regularization term is added as an incentive to minimize unnecessary coefficients. The resulting loss function is:

$$L = \frac{1}{N} \sum_{i \in I} \left(smooth_{L1}(S_i - \hat{S}_i) + \lambda \left(\sum_{k_l^m \in \mathbf{k}_i} |(k_l^m)^i| \right) \right) \tag{4}$$

where $|(k_l^m)^i|$ is the absolute value of the coefficient. The loss is used to update the MLP parameters Ψ.

To reconstruct images from the trained model, a set C of (x, y, z) coordinates is generated at the desired spatial resolution, as well as a set B of directions in the desired angular resolution. A dMRI dataset is reconstructed by first positionally encoding, and mapping every coordinate $\mathbf{c} \in C$ to $\mathbf{k_c}$ using (1), and then sampling the SH-series parameterized with $\mathbf{k_c}$ for every direction $\mathbf{b} \in B$. Effectively this applies (3) to every coordinate-direction pair, but only calculates $\mathbf{k_c}$ once for every input coordinate.

The model has the following hyperparameters: l_{max} sets the maximum degree of the SH, l_{pos} sets the number of positional encodings, σ scales the positional encoding, n_layers sets the number of layers in the MLP, $hidden\_dim$ sets the number of neurons in each layer, lr is the learning rate, λ scales the L1 regularization.

2.3 Implementation

The model is implemented in python version 3.9.16, with pytorch version 2.0.0. MRtrix3 version 3.0.4 is used to calculate DTI metrics and fODFs, and to visualize results. Scilpy[1] version 1.5 is used (with python version 3.10.10) to calculate fODF based metrics, and to create interpolated spherical functions. The BET of FSL version 6.0.6.4, is used for brain mask segmentation.

2.4 Experiments

Reconstruction and Angular Upsampling of the dMRI Signal. To assess if the proposed model can reproduce the original data, NeSH is fit on 30 gradient directions of the $b = 1000$ s/mm^2 shell. A grid-search is performed over the hyperparameters. Visual inspection of the gradient images, as well as DTI metrics and fODF glyphs, determine which settings produce the best results. Then, to assess if these settings can be applied to a different set of gradient directions, the settings found in part one are used to fit the model on 90, 60, 45, 30, 15, 10 and 3 gradient directions for both the $b = 1000$ s/mm^2 and $b = 3000$ s/mm^2 dMRI acquisitions. As a comparison, spherical harmonics interpolation (SHI) [5] is fit on the same number of gradients. The root mean squared error (RMSE) is calculated for each of the models between the input gradient images and the reconstructed gradient images it produces:

$$\sqrt{\frac{1}{WHD|B|} \sum_{x=1}^{W} \sum_{y=1}^{H} \sum_{z=1}^{D} \sum_{\mathbf{b} \in B} (S(x, y, z, \mathbf{b}) - \hat{S}(x, y, z, \mathbf{b}))^2} \tag{5}$$

---

[1] https://github.com/scilus/scilpy.

where W, H, and D are the width, height, and depth of the image, B is the set of gradient directions with size $|B|$, $S(x, y, z, \boldsymbol{b})$ is the measured signal, and $\hat{S}(x, y, z, \boldsymbol{b})$ is the reconstructed signal at location x, y, z for gradient direction $\boldsymbol{b}$. Finally the capabilities of the model to upsample in the angular domain are assessed. The resulting models from the second part are sampled in all 90 gradient directions. The RMSE is calculated between the 90 original gradient images and the 90 reconstructed gradient images using (5). In all experiments the RMSE is only calculated within a brain mask.

Spatial Upsampling. The data modeled with NeSH can be sampled in any spatial resolution. This experiment assesses the quality of the data when upsampled in spatial domain. The HCP dataset is downsampled from 1.25 mm to 2.5 mm isotropic voxels. NeSH is fit on the downsampled dataset using 90 gradient directions, and then sampled at 1.25 mm isotropic resolution. The downsampled dataset is also upsampled to the original 1.25 mm isotropic resolution using cubic interpolation. For the resulting datasets a color encoded FA map is calculated and visualized to compare the results.

DTI and fODF Metrics. In this experiment we assess if the data modeled with NeSH can be used to produce three common dMRI microstrutural metrics. Two DTI metrics: mean diffusivity (MD) and fractional anisotropy (FA), and one fODF metric: total apparent fiber density (AFD, [10]). The metrics are calculated for 90 gradients, 90 gradients reconstructed by NeSH fit on 90 gradients, 30 gradients, 90 gradients reconstructed by NeSH fit on 30 gradients, and 90 gradients reconstructed by SHI fit on 30 gradients. The $b = 1000$ s/mm^2 shell was used for the DTI metrics, and the $b = 3000$ s/mm^2 shell for AFD. The three measures are compared to the ones obtained from the full 90 gradients set by computing and visualizing a difference map.

fODF Estimation. This experiment is used to assess if fODFs can be generated from data modeled with NeSH. The same datasets as in the previous experiment are used. A response function is first extracted from the dMRI acquisitions using the single shell implementation of the algorithm by Tournier [15]. Secondly, the fODFs are calculated using single shell CSD [14]. For all five datasets, the $b = 3000$ s/mm^2 shell is used. Results are visualized by showing fODF glyphs.

3 Results

Reconstruction and Angular Upsampling of dMRI Signal. The grid-search over the parameters resulted in the following hyperparameter settings: $l_{max} = 8$ (for models trained ≤ 10 gradient directions $l_{max} = 2$), $l_{pos} = 12$ resulting in an input size of 75 (12 sine and cosine encodings for each dimension + raw coordinates), $\sigma = 4$, $n\_layers = 4$, $hidden\_dim = 2048$, $lr = 10 \times 10^{-5}$, $\lambda = 10 \times 10^{-6}$. The Adam optimizer was used with default settings. The model is trained for 5 epochs with a batch size of 1000. Figure 2 shows a slice of dMRI data for a single gradient direcion, the output generated by NeSH, as well as the mean squared error (RMSE) between the two images. NeSH produces a smoother image, removing the noise from the input image. The noise appears to be randomly distributed, without anatomical residuals.

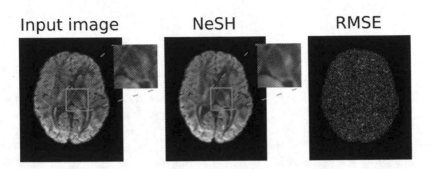

Fig. 2. A single axial slice of dMRI data, from a single gradient direction shown as baseline (column one) and as reconstructed by NeSH (column two). The root mean squared error (RMSE) between the two images is shown in column three. A two time magnification of the area in the green boxes shows the denoising effect more clearly. (Color figure online)

The comparisons of reconstruction error between NeSH and SHI are shown in Fig. 3a. NeSH has a higher RMSE when reconstructing a lower number of gradients, which lowers with an increasing number of gradients. SHI has a lower RMSE when reconstructing a lower number of gradients, which increases with an increasing number of gradients. SHI has a consistently lower RMSE compared to NeSH on both $b = 1000$ s/mm^2 and $b = 3000$ s/mm^2.

The comparisons of upsampling error between NeSH and SHI are shown in Fig. 3b. For both Nesh and SHI the RMSE of the upsampled data lowers when the model is fit on more gradient directions. SHI has a lower RMSE for all gradient subsets for $b = 1000$ s/mm^2, and for all subsets with more then 15 gradients for $b = 3000$ s/mm^2.

Spatial Upsampling. Figure 4 shows the color encoded FA maps for this experiment. The dataset reconstructed by NeSH fit on 90 gradients is able to reconstruct details in the cerebellar cortex and cerebellar white matter that are lost in cubic interpolation. Finer-grained details of the 1.25 mm isotropic voxel data are lost in both upsampling methods.

DTI Metrics and fODF Metrics. Figure 5 shows the results of this experiment. The DTI metrics (MD, FA) show low error when downsampling, and in NeSH and SHI reconstructions on both 90 and 30 gradients, when compared to

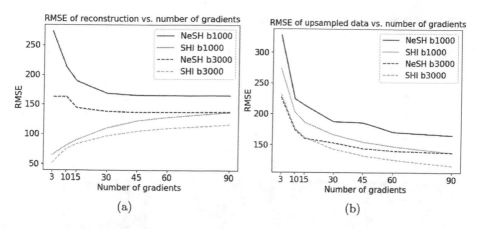

Fig. 3. Root mean squared error (RMSE) of the reconstructed dMRI volumes by NeSH and spherical harmonics interpolation (SHI) compared to the gradient images used to fit the model (a) and compared to the full set of 90 gradient images (b), for both $b = 1000$ s/mm^2 and $b = 3000$ s/mm^2.

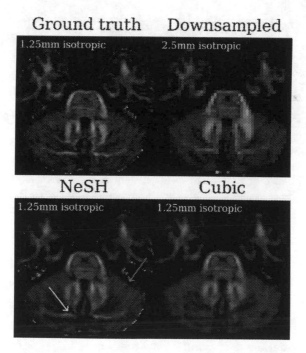

Fig. 4. A slice of color encoded FA maps at the cerebellar level, generated for different datasets. Clockwise starting top left: the 1.25 mm isotropic original image (ground truth), the downsampled 2.5 mm isotropic image, 1.25 mm isotropic image upsampled using cubic interpolation, 1.25 mm isotropic image upsampled using NeSH. The arrows show two areas where NeSH is able to reconstruct details of the cerebellum which are lost in cubic interpolation.

the metrics calculated on 90 gradients. Downsampling and SHI both over- and underestimate the metric, with most errors located in the white matter areas. NeSH more frequently underestimates the metrics and the errors are located more in the grey matter areas. The AFD is overestimated by NeSH reconstructions, and underestimated when downsampling or resconstructing using SHI. In AFD the distribution of error is similar for all methods, but downsampling and SHI show an underestimation of the AFD, while NeSH shows errors in both directions.

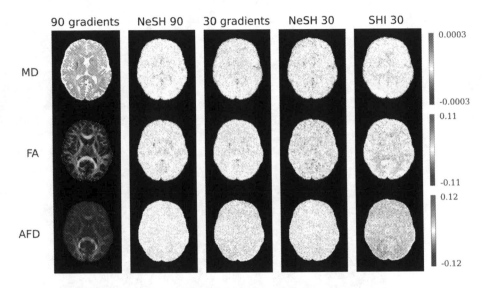

Fig. 5. Visualization of mean diffusivity (MD), fractional anisotropy (FA) and total apparant fiber density (AFD). In the first column the metrics are shown as a map for a single axial slice of the volume when calculated on the full set of 90 gradient images. The remaining columns show the difference map for the other datasets when compared to the 90 gradient images. Blue signifies negative difference, red signifies positive difference. (Color figure online)

fODF Estimation. The visualization of a group of fODFs in the centrum semi-ovale and the descending part of the CST can be seen in Fig. 6. The glyphs created from the NeSH reconstructions show a smooth, structurally-coherent change, while maintaining the important information, i.e. the crossing of fibers in the centrum semi-ovale. In presence of a big fiber tract such as the CST, the NeSH reconstruction shows a decrease in amplitude in other directions. In other methods, the fODFs exhibit more noise and less alignment between voxels, while the peaks appear sharper.

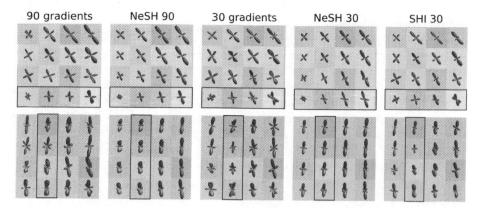

Fig. 6. Magnified coronal view showing the fiber orientation distributions glyps for the different datasets on a background of a T1-weighted image. First row: centrum semiovale, highlighted region shows the increased intervoxel consistency in NeSH modelled data. Second row: descending part of the corticospinal tract, highlighted region shows increased intervoxel consistency in NeSH modelled data, as well as a decrease in the size of the crossing fibers.

4   Discussion

We have introduced a novel method to model only a single acquisition of dMRI data using a neural representation of spherical harmonics, called NeSH. We show that dMRI data reconstructed by NeSH appears to be denoised compared to the original data (Fig. 2). We hypothesise that the model is able to capture the continuous structures of the brain, but not the erratic nature of the noise. The RMSE lies consistently higher for both the reconstruction and upsampling compared to SHI (Fig. 3), which could partly be explained by the removal of noise. For the reconstruction of the input gradients, NeSH performs worse with a decreasing number of input gradients. Possibly this indicates that to find a good representation NeSH needs a minimum number of gradients, which appears to be around 15. Both NeSH and SHI show an increase in RMSE when upsampling to 90 gradients from a decreasing number of gradients, as is to be expected.

We also show that NeSH can also be used to upsample in the spatial domain. Figure 4 shows that NeSH is able to reconstruct details that are not clearly visible in the 2.5 mm isotropic data, but are present in the 1.25 mm ground truth. This strengthens the hypothesis that using multiple gradient directions, NeSH can model a continuous representation of the dMRI data. While the achievable level of detail is lower than achieved by Alexander et al. [2], it does not rely on a prior learned from a large dataset.

Furthermore, we show that the reconstructed dMRI volumes can be used to calculate MD, FA, and AFD. Compared to the metrics calculated using 90 gradient direction, NeSH differs mostly in the gray matter areas, while downsampled volumes and SHI reconstructed volumes have differences in the white matter areas. This supports our hypothesis that NeSH benefits from the structural continuity of the fiber bundles to model the data. As the white matter areas are usually the areas of interest, this could be seens as a benefit of NeSH. The increased brightness in the posterior commissure and surrounding tissue can be explained by the lack of bias field correction in pre-processing.

Finally, we show that fODFs generated from NeSH reconstructions have a smooth change in fiber directions between voxels. This is also supportive of the structural continuity hypothesis. The 90 gradient, 30 gradient, and SHI reconstructed FODs show a more erratic pattern, in which the FODs are less aligned overall. The decrease in size of the crossing fibers in the descending part of the CST shows that NeSH prioritizes the major bundle in this area. Some information on possible smaller bundles is now lost, however, which is something that should be looked into in future versions.

Limitations. The lack of a gold-standard in dMRI complicates the interpretation of the experiments. The denoising effect shown in Fig. 2 is an example. We cannot be certain if the representation modelled by NeSH is a more realistic one than the more noisy representation or just a smoother one. In the last experiment NeSH consistently shows lower values in grey matter areas. The fODFs in the grey matter area correspondingly show a less peaked, smaller amplitude. Compared to existing techniques this can be interpreted as an error, but anatomically it makes sense as there are no large fiber bundles in the grey matter. Additionally, with no ground truth data, it is difficult to assess how good the representation outside of the voxels actually is. Synthetic datasets with known ground truths can provide a better idea. Furthermore, both the architecture and positional encodings used in this paper are simple. Many developments in the field NeRF have taken place since [12,20]. Architectural and methodological changes to NeSH could lead to further improvement. Finally, we choose to model the dMRI signal directly through an SH-series, in order to evaluate the data quality with a variety of downstream tasks. This is not a necessity. Anything that can be transformed into dMRI signal can be modelled by NeSH (e.g. peak directions or fODFs, which can be convolved into a diffusion weighted signal).

Future Work. Future work will further investigate the advantages of modelling dMRI data in a continuous space, as well as further evaluate the findings of the experiments. First, the quality and usability of the denoising properties of NeSH should be compared to other existing denoising methods. Second, using clinical datasets of lower angular and spatial resolution can provide insight into the 'real-world' clinical applicability of NeSH. This is especially interesting in MRI acquisitions of pathology (e.g. a glial-cell tumor) in the brain, as models relying on a prior learned on outside data might fail here. The harmonization of dMRI datasets across scanners and protocols [9,13], is another area of research where NeSH can be applied. Third, fiber tracking is a common use-case for dMRI. We have performed fibertracking using iFOD2 [16] with tract masks and begin- and endpoint inclusion for different numbers of gradient directions. This showed no major differences between the different methods for all inspected tracts. An explanation for this is the high spatial resolution of the HCP data, which allows tracts to be generated even with a downsampled angular resolution. Further research on datasets with lower spatial resolution will have to show the value of using NeSH reconstructions for fiber tracking. Fourth, a recent paper by Mancini et al. [7] has shown how compression of dMRI data using sinusoidal representation networks (SIRENs) [11] does not lead to reduced quality in downstream tasks. Using a SIREN architecture could also prove useful for the SH-based approach we have described. Finally, the generalization of the model to other subjects, protocols, and scanners has to be evaluated. We have performed a preliminary experiment which showed comparable results for signal reconstruction and fODFs.

5   Conclusion

Modeling dMRI data using NeSH produces results in downstream tasks with similar or possibly better results than established methods. It also shows promising results in the field of angular and spatial upsampling. NeSH can make use of the structural coherence in the brain, and does not rely on a prior learned on other datasets. The experiments in this paper provide an interesting avenue for modeling dMRI data, which should be further explored in future research.

References

1. Aja-Fernandez, S., et al.: Validation of deep learning techniques for quality augmentation in diffusion MRI for clinical studies (2023)
2. Alexander, D.C., et al.: Image quality transfer and applications in diffusion MRI. Neuroimage **152**, 283–298 (2017)
3. Basser, P.J., Mattiello, J., LeBihan, D.: Estimation of the effective self-diffusion tensor from the NMR spin echo. J. Magn. Reson. Ser. B **103**(3), 247–254 (1994)
4. Bonet-Carne, E., Grussu, F., Ning, L., Sepehrband, F., Tax, C.M.: Computational Diffusion MRI: International MICCAI Workshop, Granada, Spain, September 2018. Springer, Cham (2019). https://doi.org/10.1007/978-3-030-05831-9
5. Descoteaux, M., Angelino, E., Fitzgibbons, S., Deriche, R.: Regularized, fast, and robust analytical q-ball imaging. Magn. Reson. Med.: Off. J. Int. Soc. Magn. Reson. Med. **58**(3), 497–510 (2007)
6. Girshick, R.: Fast R-CNN. In: Proceedings of the IEEE International Conference on Computer Vision, pp. 1440–1448 (2015)
7. Mancini, M., Jones, D.K., Palombo, M.: Lossy compression of multidimensional medical images using sinusoidal activation networks: an evaluation study. In: Cetin-Karayumak, S., et al. (eds.) CDMRI 2022. LNCS, vol. 13722, pp. 26–37. Springer, Cham (2022). https://doi.org/10.1007/978-3-031-21206-2_3
8. Mildenhall, B., Srinivasan, P.P., Tancik, M., Barron, J.T., Ramamoorthi, R., Ng, R.: NeRF: representing scenes as neural radiance fields for view synthesis. Commun. ACM **65**(1), 99–106 (2021)
9. Ning, L., et al.: Cross-scanner and cross-protocol multi-shell diffusion MRI data harmonization: algorithms and results. Neuroimage **221**, 117128 (2020)

10. Raffelt, D., et al.: Apparent fibre density: a novel measure for the analysis of diffusion-weighted magnetic resonance images. Neuroimage **59**(4), 3976–3994 (2012)
11. Sitzmann, V., Martel, J.N.P., Bergman, A.W., Lindell, D.B., Wetzstein, G.: Implicit neural representations with periodic activation functions (2020)
12. Tancik, M., et al.: Fourier features let networks learn high frequency functions in low dimensional domains. Adv. Neural. Inf. Process. Syst. **33**, 7537–7547 (2020)
13. Tax, C.M., et al.: Cross-scanner and cross-protocol diffusion MRI data harmonisation: a benchmark database and evaluation of algorithms. Neuroimage **195**, 285–299 (2019)
14. Tournier, J.D., Calamante, F., Connelly, A.: Robust determination of the fibre orientation distribution in diffusion MRI: non-negativity constrained super-resolved spherical deconvolution. Neuroimage **35**(4), 1459–1472 (2007)
15. Tournier, J.D., Calamante, F., Connelly, A.: Determination of the appropriate b value and number of gradient directions for high-angular-resolution diffusion-weighted imaging. NMR Biomed. **26**(12), 1775–1786 (2013)
16. Tournier, J.D., Calamante, F., Connelly, A., et al.: Improved probabilistic streamlines tractography by 2nd order integration over fibre orientation distributions. In: Proceedings of the International Society for Magnetic Resonance in Medicine, vol. 1670. Ismrm (2010)
17. Tournier, J.D., et al.: MRtrix3: a fast, flexible and open software framework for medical image processing and visualisation. NeuroimageD **202**, 116137 (2019)
18. Van Essen, D.C., et al.: The WU-Minn human connectome project: an overview. Neuroimage **80**, 62–79 (2013)
19. Yu, A., Li, R., Tancik, M., Li, H., Ng, R., Kanazawa, A.: Plenoctrees for real-time rendering of neural radiance fields. In: Proceedings of the IEEE/CVF International Conference on Computer Vision, pp. 5752–5761 (2021)
20. Zhu, F., Guo, S., Song, L., Xu, K., Hu, J., et al.: Deep review and analysis of recent NeRFs. APSIPA Trans. Signal Inf. Process. **12**(1) (2023)

A Unified Learning Model for Estimating Fiber Orientation Distribution Functions on Heterogeneous Multi-shell Diffusion-Weighted MRI

Tianyuan Yao[1(✉)], Nancy Newlin[1], Praitayini Kanakaraj[1], Vishwesh Nath[3],
Leon Y. Cai[1], Karthik Ramadass[1], Kurt Schilling[2], Bennett A. Landman[1],
and Yuankai Huo[1]

[1] Vanderbilt University, Nashville, TN 37215, USA
tianyuan.yao@vanderbilt.edu
[2] Vanderbilt University Medical Center, Nashville, TN 37215, USA
[3] NVIDIA Corporation, Santa Clara and Bethesda, USA

Abstract. Diffusion-weighted (DW) MRI measures the direction and scale of the local diffusion process in every voxel through its spectrum in q-space, typically acquired in one or more shells. Recent developments in micro-structure imaging and multi-tissue decomposition have sparked renewed attention to the radial b-value dependence of the signal. Applications in tissue classification and micro-architecture estimation, therefore, require a signal representation that extends over the radial as well as angular domain. Multiple approaches have been proposed that can model the non-linear relationship between the DW-MRI signal and biological microstructure. In the past few years, many deep learning-based methods have been developed towards faster inference speed and higher inter-scan consistency compared with traditional model-based methods (e.g., multi-shell multi-tissue constrained spherical deconvolution). However, a multi-stage learning strategy is typically required since the learning process relies on various middle representations, such as simple harmonic oscillator reconstruction (SHORE) representation. In this work, we present a unified dynamic network with a single-stage spherical convolutional neural network, which allows efficient fiber orientation distribution function (fODF) estimation through heterogeneous multi-shell diffusion MRI sequences. We study the Human Connectome Project (HCP) young adults with test-retest scans. From the experimental results, the proposed single-stage method outperforms prior multi-stage approaches in repeated fODF estimation with shell dropoff and single-shell DW-MRI sequences.

Keywords: DW-MRI · multi-shell Deep learning

1 Introduction

Diffusion-weighted magnetic resonance imaging (DW-MRI) is essential for the non-invasive reconstruction of the microstructure of the human *in vivo* brain [2,10,27].

© The Author(s), under exclusive license to Springer Nature Switzerland AG 2023
M. Karaman et al. (Eds.): CDMRI 2023, LNCS 14328, pp. 13–22, 2023.
https://doi.org/10.1007/978-3-031-47292-3_2

Substantial efforts have shown that other advanced approaches can recover more elaborate reconstruction of the microstructure [7,15,20] and these methods are collectively referred to as high angular resolution diffusion imaging (HARDI). HARDI methods have been broadly proposed in two categories of single-shell acquisitions and multi-shell acquisitions (i.e., using multiple b-values). A majority of single-shell HARDI methods utilize spherical harmonics (SH) based modeling as in q-ball imaging (QBI) [25], constrained spherical deconvolution (CSD) [24], and many others. However, SH-based modeling cannot directly leverage additional information provided by multi-shell acquisitions as the SH transformation does not allow for a representation of the radial complexity that is introduced by different b-value. SH has been combined with other bases to represent multi-shell data, e.g., solid harmonics [8], simple harmonic oscillator reconstruction (SHORE) [5], and spherical polar Fourier imaging [4].

Deep learning (DL) has revolutionized many different domains in medical imaging [23], and DW-MRI parameter estimation is no different. Lots of DW-MRI methods have been developed that utilize the powerful data-driven capabilities of deep learning, yielding improved accuracy over conventional fitting when the acquisition scheme has a limited number of measurements [17,28]. However, most methods are only focused on the translation of single-shell data to DW-MRI parameters, and in contrast, the multi-shell methods get neglected due to the complexity associated with multi-shell data [12,18]. Moreover, the SHORE-based DL methods typically used a multi-stage design [19]. For instance, the algorithm must first optimize a specific optimal SHORE representation and then optimize the fiber orientation distribution function (fODF) estimation. Such methods are prone to overfitting, lower inference time, and complicated parameter tuning.

As shown in Fig. 1, in this paper, we propose a single-stage dynamic network with both the q-space and radial space signal based on a spherical convolutional neural network. We evaluated the resultant representation by targeting it to multi-shell multi-tissue CSD (MSMT-CSD). Both fiber orientation estimation and recovery of tissue volume fraction are evaluated. The contribution of this paper is three-fold:

- We proposed a unified dynamic network with the single-stage spherical convolutional neural network that can recover/predict microstructural measures.
- The proposed method is universally applicable to perform learning-based fODF estimation using a single deep model for various combinations of multiple shells.
- The proposed method achieved an overall superior performance compared with model-based and data-driven benchmarks.

2   Related Work

2.1   Multi-shell Multi-tissue Constrained Spherical Deconvolution

Multi-Shell Multi-Tissue Constrained Spherical Deconvolution (MSMT-CSD) [15] is a technique developed to overcome the limitations of traditional single-shell diffusion MRI methods, which are unable to resolve the complex fiber orientations of multiple tissue types in the brain. MSMT-CSD is able to separate the contribution of different tissue types (such as gray matter, white matter, and cerebrospinal fluid) to the diffusion signal by modeling the diffusion signal as a combination of multiple shells with different b-values. This modeling-based method has been a conventional method for multi-tissue micro-architecture estimation.

2.2 Learning-Based Estimation

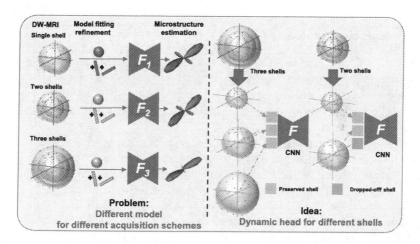

Fig. 1. Utilizing multi-shell DW-MRI signals in deep learning usually requires independent models trained for each specific shell configuration as conventional SH-based modeling cannot directly leverage additional information (radial space) provided by multi-shell acquisitions. In our study, the dynamic head aims to improve the network expressiveness by learning and adaptively adjusting the first convolution layer for different shell configurations.

Recently, machine learning (ML) and deep learning (DL) techniques have demonstrated their remarkable abilities in neuroimaging. Such approaches have been applied to the task of microstructure estimation [19], aiming to directly learn the mapping between input DW-MRI scans and output fiber tractography [3, 21] while maintaining the necessary biological characteristics and reproducibility for clinical translation. Such studies have illustrated that DL is a promising tool that uses nonlinear transforms to extract features from high-dimensional data. Data-driven approaches can be useful in validating the hypothesis of the existence of untapped information because they generalize toward the ground truth.

3   Methods

3.1   Preliminaries

Traditional deep learning frameworks are not generalizable to new acquisition schemes. This complicates the application of a DL model to data acquired from multiple sites.

Our model aims to train a DL framework that can be adapted to an arbitrary number of available multi-shell DW-MRI sequences. To serve this motivation, we employ a dynamic head (DH) design to handle the multi-shell problem on the three most common b-values: 1000, 2000, and 3000 s/mm^2. Additionally, to tackle the problem of a varying number of gradient directions on each shell (b-value), we leverage the spherical CNNs with the traditional 'modeling then feeding to a fully connected network (FCN)' strategy. In this study, we employ the fODF estimation as our chosen task to perform assessments on different methods.

3.2   Dynamic Head Design

A dynamic head design in multi-modality deep learning [16] offers a flexible way to handle diverse data types within a single model, adapting its behavior to best suit the input modality. In our scenario, we intend to use a dynamic head that allows the neural network to effectively deal with diverse inputs from different shells by adapting its processing mechanism accordingly.

Note that with K shells in our scheme, there are 2^K - 1 configurations. To improve the network expressiveness, we devise a dynamic head to adaptively generate model parameters conditioned on the availability of input shells. We use a binary code $m \in \mathbb{R}^K$ indicates that m is a vector with K real-valued entries or components. $K \in [0, 1]$ that 0/1 represent the absence/presence of each shell. To mitigate the large input variation caused by artificially zero-ed channels, we use the dynamic head to generate the parameters for the first convolutional layer.

3.3   Spherical Convolution

To extract features from DW-MRI signals, the first and most common deep learning network architecture applied to dMRI is the fully connected network(FCN) [1, 18], Conventionally these have been implemented the following:

$$y = F_{FCN}(x|\theta_{FCN}) \tag{1}$$

where F_{FCN} is a fully-connected network with trainable parameters θ_{FCN}, has signal input x and y is the ground truth dMRI parameters. Given a loss function, L tailored for a specific downstream task and the function is learned by optimizing the trainable parameters θ_{FCN} and can be expressed as follows:

$$\tilde{\theta}_{FCN} = \arg \min_{\theta_{FCN}} L(yi, F(xi|\theta_{FCN})) \tag{2}$$

The dMRI signal xi serves as the ith input for the network with corresponding ground truth output yi, and it does not consider the acquisition information, making the network unaware of the acquisition scheme. This lack of knowledge poses an issue when incorporating new data acquired at a different location with a distinct acquisition scheme. The accuracy of estimation from a new set of DW-MRIs depends on the consistency of the acquisition scheme with the training set. Additionally, the FCN's design does not account for rotational equivariance, which could result in requiring a varied range of tissue microstructure orientations in the training dataset for accurate estimation independent of fiber orientation.

Theoretically, Spherical CNNs offer an advantage over FCNs regarding both the robustness of the gradient scheme and the distribution of training data [11,22]. The Spherical CNN's architecture differs from FCNs, but not in the conventional sense. Instead of convolution across multiple voxels, Spherical CNNs perform convolution over the spherical image space. Hence, like FCNs, they are voxelwise networks. At each voxel, the spherical image is created from the dMRI signals and their corresponding gradient scheme. This architecture can naturally address the limitations of FCNs in two ways. Firstly, unlike FCNs, Spherical CNNs inherently recognize the gradient scheme present in their input, as illustrated by the following equation:

$$y = F_{S-CNN}(x, G | \theta_{S-CNN}) \tag{3}$$

Here, F_{S-CNN} represents the Spherical CNNs, characterized by their trainable parameters θ_{S-CNN}. One of the distinct advantages of Spherical CNNs is their explicit consideration of the gradient scheme in the input. This capability enables them to adeptly manage variations in gradient schemes that may arise from different acquisition protocols or disparate imaging sites. Moreover, owing to their inherent spherical structure, Spherical CNNs can more effectively handle the distribution of training data that resides in a spherical domain. Cumulatively, the unique attributes of spherical convolution present significant improvements in the accuracy and robustness when analyzing diffusion MRI signals. During the training phase, the shared network F_{S-CNN} utilizes the input data. This data comprises $2^K - 1$ distinct shell configurations, which can be described as

$$\tilde{x}^k = \delta^k x^k, (k \in 1, ..., K) \tag{4}$$

where δ^k is a Bernoulli selector variable that can take on values in 0, 1. By aiming at diverse diffusion properties denoted by y and combined with the dynamic head setting, the learning objective at the ith input can be articulated as:

$$\tilde{\theta}_{S-CNN} = \arg \min_{\theta_{S-CNN}} L(yi, F(\tilde{x}^k i, Gi | \theta_{S-CNN})) \tag{5}$$

4   Experiments

4.1   Data and Implementation Details

We have chosen DW-MRI from the Human Connectome Project - Young Adult (HCP-ya) dataset [10,26], 45 subjects with the scan-rescan acquisition were used (a total of 90 images). The acquisitions had b-values of 1000, 2000, 3000 s/mm^2 with 90 gradient directions on each shell. A T1 volume of the same subject was used for WM segmentation using SLANT [13]. All HCP-ya DW-MRI was distortion corrected with top-up and eddy [14]. 30 subjects are used as training data while 10 subjects were used as evaluation and 5 subjects as testing data.

We performed shell extraction on all the data. Every subject has seven different shell configurations which are the permutations of all three b-values $\{\{1K, 2K, 3K\}$, $\{1K, 2K\}$, $\{2K, 3K\}$, $\{1K, 3K\}$, $\{1K\}$, $\{2K\}$, $\{3K\}\}$. Ground-truth fODF maps were computed from MSMT-CSD using the DIPY library with the default settings [9]. 8th order SH were chosen for data representation with the 'tournier07' basis [24]. The white matter fODF and the volume fraction which refers to the proportion of the volume of the voxel that is occupied by each tissue type, are combined together as the targeted sequence.

Inspired by Nath et al. [19], we employed the simple harmonic oscillator-based reconstruction and estimation (SHORE) as another baseline representation. SHORE modeling is known to capture the complex diffusion signal across different b-values without resorting to multi-compartment models, where the SHORE basis function is given by $Z_{nlm}(q, \Theta) = R_n(q)Y_{lm}(\Theta)$. As for the single shell dMRI signal, $|q|$ is constant, and the variability in $E(\mathbf{q})$ is primarily captured by $Y_{lm}(\Theta)$. The richness of the model (i.e., maximum order N) is likely needed to be limited for single-shell data to avoid overfitting. Thus, the 6th radial order SHORE basis is employed as a baseline representation for both single-shell and multi-shell dMRI signals to fit the fiber ODF. The SHORE scaling factor ζ defined in units of mm^{-2} as $\zeta = 1/8\pi^2\tau MD$ is calculated based on the mean diffusivity (MD) obtained from the data. Given that both SHORE base signal ODF and SH base fiber ODF have the same underlying information. We apply deep learning to map the intricate relationships and patterns from one representation to another.

4.2 Experimental Setting

We first trained separate models for each shell configuration. The models consist of four fully connected layers with ReLU activation function. The number of neurons per layer is 400, 48, 200, and 48. The input is the 1×50 vector of the shore basis signal ODF, and the output is the combination 1×45 vectors of the SH basis WM fODF and the 1×3 vector of tissues fraction. The models are then tested on the different shell configurations. By simply feeding all the shell configuration data (all labeled with reconstructed fODF from data with all shells) to the FCN as a baseline 'unified' deep learning model.

We assess the impact of dynamic head strategy by evaluating the performance of the unified models against independent models trained for each specific shell configuration. Furthermore, the generalizability of the different representations with dynamic head designs was assessed. For the spherical convolution, we used an architecture known as the hybrid spherical CNN as described in [6]. The architecture consists of a S^2 convolutional layer and a $SO(3)$ convolutional layer and is followed by three channel-wise activations and two restricted generalized convolutions until the final restricted generalized convolution maps down to a rotationally invariant representation. The specific network parameters follow the spherical MNIST experiment [6]. The diffusion signals from different shells are 1-to-1 densely sampled to map between six directional dMRI signals and the 6 independent values of the diffusion tensor. After the rotational invariant features are extracted. They are concatenated and fed into fully connected layers(the same hidden size as above) which perform the final estimation.

4.3 Evaluation Metric

To evaluate the predictions from the proposed methods, we calculated the mean squared error of the volume fractions with ground truth sequences. Then we compute the angular correlation coefficient (ACC, Eq. 6) between the predicted fODF and the ground truth fODF over the white matter region. ACC is a generalized measure for all fiber population scenarios. It assesses the correlation of all directions over a spherical harmonic expansion. In brief, it provides the estimate of how closely a pair of fODFs are related on a scale of -1 to 1, where 1 is the best measure. Here 'u' and 'v' represent sets of SH coefficients.

$$ACC = \frac{\sum_{k=1}^{L}\sum_{m=-k}^{k}(u_{km})(v_{km}^{*})}{[\sum_{k=1}^{L}\sum_{m=-k}^{k}|u_{km}|^{2}]^{0.5} \cdot [\sum_{k=1}^{L}\sum_{m=-k}^{k}|v_{km}|^{2}]^{0.5}} \tag{6}$$

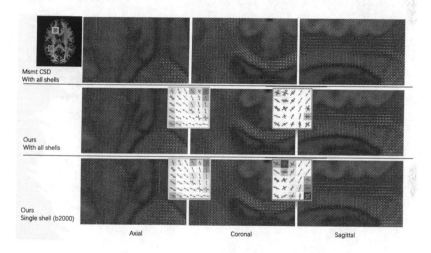

Fig. 2. This is a visualization of the fODF prediction and the correlation with the GT in different views. The background of the zoom-in patches shows the ACC spatial map with the GT signals.

5 Experimental Results

We compared the performances of the unified models against independent models trained for each specific shell configuration. A qualitative result of fODF predictions and GT are shown in Fig. 2. As shown in Table 1, the independent models that are trained are thus more likely to outperform others in their own shell configuration and these models can be considered as the upper bounds for each shell configuration. With the dynamic head settings, the unified model with spherical convolution outperforms the other models in the single shell configuration. Additionally, the ACC is a sensitive generalized metric, the performances need further evaluation. We assessed how good our prediction was by evaluating the scan/rescan consistency and volume fraction prediction Table 2.

Table 1. Performances of the unified models against independent models in different shell configurations

Model	C_1	C_2	C_3	$C_{1,2}$	$C_{2,3}$	$C_{1,3}$	$C_{1,2,3}$	Ave.
M_1	0.808	0.725	0.732	0.752	0.734	0.751	0.788	0.756
M_2	0.762	0.815	0.756	0.744	0.749	0.745	0.774	0.764
M_3	0.757	0.724	0.814	0.734	0.753	0.760	0.779	0.760
$M_{1,2}$	0.745	0.732	0.743	0.831	0.788	0.778	0.789	0.772
$M_{2,3}$	0.734	0.744	0.738	0.802	0.825	0.786	0.786	0.774
$M_{1,3}$	0.737	0.745	0.745	0.785	0.793	0.832	0.784	0.774
$M_{1,2,3}$	0.752	0.734	0.742	0.762	0.756	0.772	0.853	0.767
All Data Feeding	0.789	0.793	0.794	0.801	0.799	0.803	0.814	0.799
DH w. SHORE [5]	0.782	0.788	0.784	0.823	0.817	0.825	0.843	0.809
DH w. SH	0.805	0.809	0.814	0.818	0.812	0.812	0.832	0.815
DH w. SC (Ours)	0.816	0.82	0.816	0.827	0.828	0.824	0.837	0.824

M_i, where i $\in$ [1, 2, 3] indicates the model is only trained on that shell configuration. C_i indicates the testing data in that shell configuration. The best and second best performances are denoted by the red mark and blue mark. The average metrics of ACC are listed in the last column

Table 2. FODF prediction assessment

Model	Shell configuration	Tissue proportion prediction	Scan-rescan consistency
Single model	1K	8.45E-04	0.862
	2K	7.92E−04	0.865
	3K	8.63E−04	0.857
	1K, 2K	7.32E−04	0.856
	2K, 3K	7.49E−04	0.86
	1K, 3K	8.02E−04	0.862
	1K, 2K, 3K	6.38E−04	0.865
DH w. SC	1K	7.27E−04	0.855
	2K	7.12E−04	0.86
	3K	7.35E−04	0.858
	1K, 2K	7.01E−04	0.858
	2K, 3K	6.79E−04	0.86
	1K, 3K	6.82E−04	0.864
	1K, 2K, 3K	5.92E−04	0.861
Silver standard: MSMT-CSD [15]			0.856

Reconstruction results from msmt-CSD are applied as silver standard in the evaluation. Wilcoxon signed-rank test is applied as a statistical assessment for scan-rescan consistency evaluation. It has a significant difference ($p < 0.001$) compared with WM fODF. The MSE is reported for evaluation of VF predictions. The ACC between scan/rescan DW-MRI over WM regions is reported.

6 Conclusion

In this paper, we propose a single-stage dynamic network with both the q-space and radial space signal based on a spherical convolutional neural network. Integrating dynamic head and spherical convolution removes the need to retrain a new network for a known b-value of DW-MRI. Besides, adjusting the last multi-layer regression network to different targets, this plug-and-play design of our method is potentially applicable to a wider range of diffusion properties in neuroimaging.

References

1. Aliotta, E., Nourzadeh, H., Sanders, J., Muller, D., Ennis, D.B.: Highly accelerated, model-free diffusion tensor MRI reconstruction using neural networks. Med. Phys. **46**(4), 1581–1591 (2019)
2. Basser, P.J., Mattiello, J., LeBihan, D.: Estimation of the effective self-diffusion tensor from the NMR spin echo. J. Magn. Reson. Ser. B **103**(3), 247–254 (1994)
3. Cai, L.Y., et al.: Convolutional-recurrent neural networks approximate diffusion tractography from T1-weighted MRI and associated anatomical context. bioRxiv, pp. 2023–02 (2023)
4. Cheng, J., Ghosh, A., Jiang, T., Deriche, R.: Model-free and analytical EAP reconstruction via spherical polar Fourier diffusion MRI. In: Jiang, T., Navab, N., Pluim, J.P.W., Viergever, M.A. (eds.) MICCAI 2010, Part I. LNCS, vol. 6361, pp. 590–597. Springer, Heidelberg (2010). https://doi.org/10.1007/978-3-642-15705-9_72
5. Cheng, J., Jiang, T., Deriche, R.: Theoretical analysis and practical insights on EAP estimation via a unified HARDI framework. In: MICCAI Workshop on Computational Diffusion MRI (CDMRI) (2011)
6. Cobb, O.J., et al.: Efficient generalized spherical CNNs. arXiv preprint arXiv:2010.11661 (2020)
7. Descoteaux, M.: High angular resolution diffusion imaging (HARDI). Wiley Encycl. Electr. Electron. Eng. 1–25 (1999)
8. Descoteaux, M., Deriche, R., Le Bihan, D., Mangin, J.F., Poupon, C.: Multiple Q-shell diffusion propagator imaging. Med. Image Anal. **15**(4), 603–621 (2011)
9. Garyfallidis, E., et al.: Dipy, a library for the analysis of diffusion MRI data. Front. Neuroinform. **8**, 8 (2014)
10. Glasser, M.F., et al.: The minimal preprocessing pipelines for the human connectome project. Neuroimage **80**, 105–124 (2013)
11. Goodwin-Allcock, T., McEwen, J., Gray, R., Nachev, P., Zhang, H.: How can spherical CNNs benefit ml-based diffusion MRI parameter estimation? In: Cetin-Karayumak, S., et al. (eds.) CDMRI 2022. LNCS, vol. 13722, pp. 101–112. Springer, Cham (2022). https://doi.org/10.1007/978-3-031-21206-2_9
12. Hansen, C.B., et al.: Contrastive semi-supervised harmonization of single-shell to multi-shell diffusion MRI. Magn. Reson. Imaging **93**, 73–86 (2022)
13. Huo, Y., et al.: 3D whole brain segmentation using spatially localized atlas network tiles. Neuroimage **194**, 105–119 (2019)
14. Jenkinson, M., Beckmann, C.F., Behrens, T.E., Woolrich, M.W., Smith, S.M.: FSL. Neuroimage **62**(2), 782–790 (2012)
15. Jeurissen, B., Tournier, J.D., Dhollander, T., Connelly, A., Sijbers, J.: Multi-tissue constrained spherical deconvolution for improved analysis of multi-shell diffusion MRI data. Neuroimage **103**, 411–426 (2014)

16. Liu, H., et al.: ModDrop++: a dynamic filter network with intra-subject co-training for multiple sclerosis lesion segmentation with missing modalities. In: Wang, L., Dou, Q., Fletcher, P.T., Speidel, S., Li, S. (eds.) MICCAI 2022. LNCS, vol. 13435, pp. 444–453. Springer, Cham (2022). https://doi.org/10.1007/978-3-031-16443-9_43

17. Müller, P., Golkov, V., Tomassini, V., Cremers, D.: Rotation-equivariant deep learning for diffusion MRI. arXiv preprint arXiv:2102.06942 (2021)

18. Nath, V., et al.: Inter-scanner harmonization of high angular resolution DW-MRI using null space deep learning. In: Bonet-Carne, E., Grussu, F., Ning, L., Sepehrband, F., Tax, C.M.W. (eds.) MICCAI 2019. MV, pp. 193–201. Springer, Cham (2019). https://doi.org/10.1007/978-3-030-05831-9_16

19. Nath, V., et al.: Deep learning reveals untapped information for local white-matter fiber reconstruction in diffusion-weighted MRI. Magn. Reson. Imaging 62, 220–227 (2019)

20. Özarslan, E., et al.: Mean apparent propagator (map) MRI: a novel diffusion imaging method for mapping tissue microstructure. Neuroimage 78, 16–32 (2013)

21. Schilling, K.G., et al.: Fiber tractography bundle segmentation depends on scanner effects, vendor effects, acquisition resolution, diffusion sampling scheme, diffusion sensitization, and bundle segmentation workflow. Neuroimage 242, 118451 (2021)

22. Sedlar, S., Alimi, A., Papadopoulo, T., Deriche, R., Deslauriers-Gauthier, S.: A spherical convolutional neural network for white matter structure imaging via dMRI. In: de Bruijne, M., et al. (eds.) MICCAI 2021, Part III. LNCS, vol. 12903, pp. 529–539. Springer, Cham (2021). https://doi.org/10.1007/978-3-030-87199-4_50

23. Suzuki, K.: Overview of deep learning in medical imaging. Radiol. Phys. Technol. 10(3), 257–273 (2017)

24. Tournier, J.D., Calamante, F., Connelly, A.: Robust determination of the fibre orientation distribution in diffusion MRI: non-negativity constrained super-resolved spherical deconvolution. Neuroimage 35(4), 1459–1472 (2007)

25. Tuch, D.S.: Q-ball imaging. Magn. Reson. Med.: Off. J. Int. Soc. Magn. Reson. Med. 52(6), 1358–1372 (2004)

26. Van Essen, D.C., et al.: The WU-Minn human connectome project: an overview. Neuroimage 80, 62–79 (2013)

27. Van Essen, D.C., et al.: The human connectome project: a data acquisition perspective. Neuroimage 62(4), 2222–2231 (2012)

28. Xiang, T., Yurt, M., Syed, A.B., Setsompop, K., Chaudhari, A.: DDM2: self-supervised diffusion mri denoising with generative diffusion models. arXiv preprint arXiv:2302.03018 (2023)

Diffusion Phantom Study of Fiber Crossings at Varied Angles Reconstructed with ODF-Fingerprinting

Patryk Filipiak[1](✉) , Timothy M. Shepherd[1] , Lee Basler[2],
Anthony Zuccolotto[2], Dimitris G. Placantonakis[3] , Walter Schneider[4] ,
Fernando E. Boada[5] , and Steven H. Baete[1]

[1] Center for Advanced Imaging Innovation and Research (CAI²R),
Department of Radiology, NYU Langone Health, New York, NY, USA
`patryk.filipiak@nyulangone.org`
[2] Psychology Software Tools, Inc., Pittsburgh, PA, USA
[3] Department of Neurosurgery, Perlmutter Cancer Center, Neuroscience Institute,
Kimmel Center for Stem Cell Biology, NYU Langone Health, New York, NY, USA
[4] University of Pittsburgh, Pittsburgh, PA, USA
[5] Radiological Sciences Laboratory and Molecular Imaging Program at Stanford,
Department of Radiology, Stanford University, Stanford, CA, USA

Abstract. White matter fiber reconstructions based on seeking local maxima of Orientation Distribution Functions (ODFs) typically fail to identify fibers crossing at narrow angles below 45°. ODF-Fingerprinting (ODF-FP) replaces the ODF maxima localization mechanism with pattern matching, allowing the use of all information stored in ODFs. In this work, we study the ability of ODF-FP to reconstruct fibers crossing at varied angles spanning 10°–90° in physical diffusion phantoms composed of textile tubes with $0.8\,\mu$m diameter, approaching the anatomical scale of axons. Our results show that ODF-FP is able to correctly identify $80 \pm 8\%$ of the crossing fibers regardless of the crossing angle and provide the highest average reconstruction accuracy.

Keywords: Fiber reconstruction · ODF-Fingerprinting · Diffusion MRI · Narrow crossing angles

1 Introduction

Diffusion MRI (dMRI) has enabled visualization of fiber structure in brain White Matter (WM) in vivo [4,14]. However, volumes of WM tissue with fibers crossing at narrow angles below 45° are challenging to reconstruct from Diffusion Weighted Images (DWIs) due to signal sampling limitations [13]. Therefore, reconstructions are often distorted by false positive or false negative fibers [17,20].

M. Karaman et al. (Eds.): CDMRI 2023, LNCS 14328, pp. 23–34, 2023.
https://doi.org/10.1007/978-3-031-47292-3_3

In dMRI, the underlying fiber structure is typically determined from Orientation Distribution Functions (ODFs) [26] whose local maxima represent preferential directions of water self-diffusion in every voxel of a DWI. Common reconstruction techniques aim at localizing these maxima through peak finding [25, 30] or statistical inference [5, 12]. ODF-Fingerprinting (ODF-FP) [1, 9] replaces the above approaches with pattern matching, which allows it to outperform several contemporary methods in reconstruction of narrow-angled fiber crossings, as was shown recently on synthetic diffusion data [9]. In this paper, we verified the reconstruction accuracy of ODF-FP on anisotropic diffusion phantoms [21] with fiber crossings at 10, 15, 20, 30, 45, and 90°.

ODF-FP is a dictionary-based fitting algorithm. For any given ODF calculated from a DWI, ODF-FP identifies the closest match in a precomputed ODF-dictionary composed of synthetically-generated samples called ODF-fingerprints. For maximum performance, ODF-fingerprints are generated using the same method as ODFs computed from the input signal [9]. This pattern matching approach allows for using the information encapsulated in the entire shape of ODF instead of relying solely on locations of ODF local maxima.

Our study shows that ODF-FP is able to correctly identify $80 \pm 8\%$ of the crossing fibers regardless of the crossing angle. Moreover, ODF-FP provides the highest mean reconstruction accuracy ($68 \pm 9\%$) given 20-degree error tolerance level, outperforming common reconstruction methods based on ODF peak finding or statistical modeling. Our results support the hypothesis that ODF-FP has potential to visualize WM fibers crossing at challenging narrow angles in vivo.

2   Methods

2.1   Data

Diffusion Phantoms. We used two anisotropic diffusion phantoms manufactured by Psychology Software Tools (Pittsburgh, PA, USA). Both phantoms contained synthetic fibers made of textile water-filled microtubes (Taxon^{TM} technology [21]) with 0.8-μm diameter. The fibers were grouped into bundles, 10 mm long and 4 mm thick, located in the axial plane and structured as follows:

- **Phantom 1** contained two identical pairs of fiber bundles crossing at 10°, 15°, and 20° (Fig. 1a).
- **Phantom 2** contained diverse fiber configurations, out of which we considered identical pairs of fiber bundles crossing at 30°, 45°, and 90° (Fig. 1b).

Acquisition. We scanned the phantoms at 2 mm isotropic resolution, with TE/TR = 74/8000 ms, using a dMRI protocol with 60 exact same sampling directions at every b-shell ($b = 250, 1000, 2250, 4000$ s/mm^2) interleaved with 20 images at $b = 0$.

Image Preprocessing. We ran the MRtrix3 [24] implementation of the Marchenko-Pastur PCA denoising algorithm (`dwidenoise`) [28] followed by removal of Gibbs ringing (`mrdegibbs`) [16] and B1 field inhomogeneity artifacts (`dwibiascorrect ants`) [27].

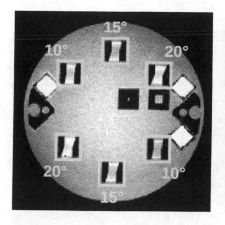 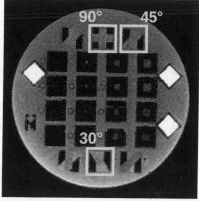

(a) Phantom 1   (b) Phantom 2

Fig. 1. Slices of the T1-weighted images of the anisotropic diffusion phantoms with the regions of interest (in yellow) containing pairs of synthetic fibers crossing at (a) 10°, 15°, 20°; (b) 30°, 45°, 90°. (Color figure online)

ODF-FP. We reconstructed diffusion ODFs [29] from the preprocessed DWIs using Radial Diffusion Spectrum Imaging (RDSI) [2] aiming to minimize spurious peaks in ODFs [2,9]. Next, we used RDSI to generate a randomized ODF-dictionary [9] with 10^5 elements containing $0 \leq N \leq 2$ crossing fibers per voxel. For this, we employed the biophysical model of WM [9,11] defined as

$$\underbrace{p_{iso}e^{-bD_{iso}}}_{\text{free water fraction}} + \sum_{i=1}^{N}\underbrace{p^{(i)}\kappa^{(i)}(b, \mathbf{g} \cdot \mathbf{n}^{(i)})}_{\text{fiber fraction}} \qquad (1)$$

where the i-th fiber fraction was modeled as

$$\kappa^{(i)}(b, \mathbf{g} \cdot \mathbf{n}^{(i)}) = f^{(i)}e^{-bD_{a,\parallel}^{(i)}(\mathbf{g}\cdot\mathbf{n}^{(i)})^2} + \left(1 - f^{(i)}\right)e^{-bD_{e,\parallel}^{(i)}(\mathbf{g}\cdot\mathbf{n}^{(i)})^2 - bD_{e,\perp}^{(i)}\left(1-(\mathbf{g}\cdot\mathbf{n}^{(i)})^2\right)}$$

$$(2)$$

with the following ranges of parameters $(i = 1, \ldots, N)$:

- volumes of free water $p_{iso} \in [0.33, 1]$ and fibers $p^{(1)}, \ldots, p^{(N)} \in [0.33, 1]$ subject to $p_{iso} + \sum_{i=1}^{N} p^{(i)} = 1$,
- intra-axonal fraction sizes $f^{(i)} \in [0, 0.5]$,
- diffusivities of free water $D_{iso} = 3 \cdot 10^{-9}$, intra-axonal $D_{a,\parallel}^{(i)} \in [0.5, 2.5] \cdot 10^{-9}$, and extra-axonal $D_{e,\parallel}^{(i)} \in [0.5, 2.5] \cdot 10^{-9}$, $D_{e,\perp}^{(i)} \in [0, 2] \cdot 10^{-9}$ $[\text{m}^2/\text{s}]$.
- fiber orientations $\mathbf{n}^{(i)} \in [-\pi, \pi] \times [-\pi/2, \pi/2]$ distributed evenly on a 321-point tessellation of a unit hemisphere, as suggested in [9].

The remaining parameters, i.e., the directions of the diffusion-encoding gradients $\mathbf{g} \in \mathbb{R}^3$ and the b-values, matched the dMRI acquisition protocol. To avoid WM model overfitting and generation of spurious fibers, we set the penalty factor $\lambda = 1 \cdot 10^{-4}$ in the ODF-FP matching mechanism [9].

The ODF-dictionary generation procedure took $20\,\mathrm{s}$ and used < 1 GB of operating memory. For simplicity, we did not consider variants with $N > 2$ fibers per voxel to highlight the aspect of reconstructions of two fibers crossing at varied angles.

2.2 Evaluation

In each phantom scan, we selected 2 axial slices intersecting with the 4-mm-thick fiber bundles. Next, we manually drew Regions Of Interest (ROIs) covering the fiber crossing volumes (Fig. 2).

To quantify our results, we first identified the numbers of fibers reconstructed in every voxel within the crossing volumes, i.e., whether 2, 1, or 0 fibers were

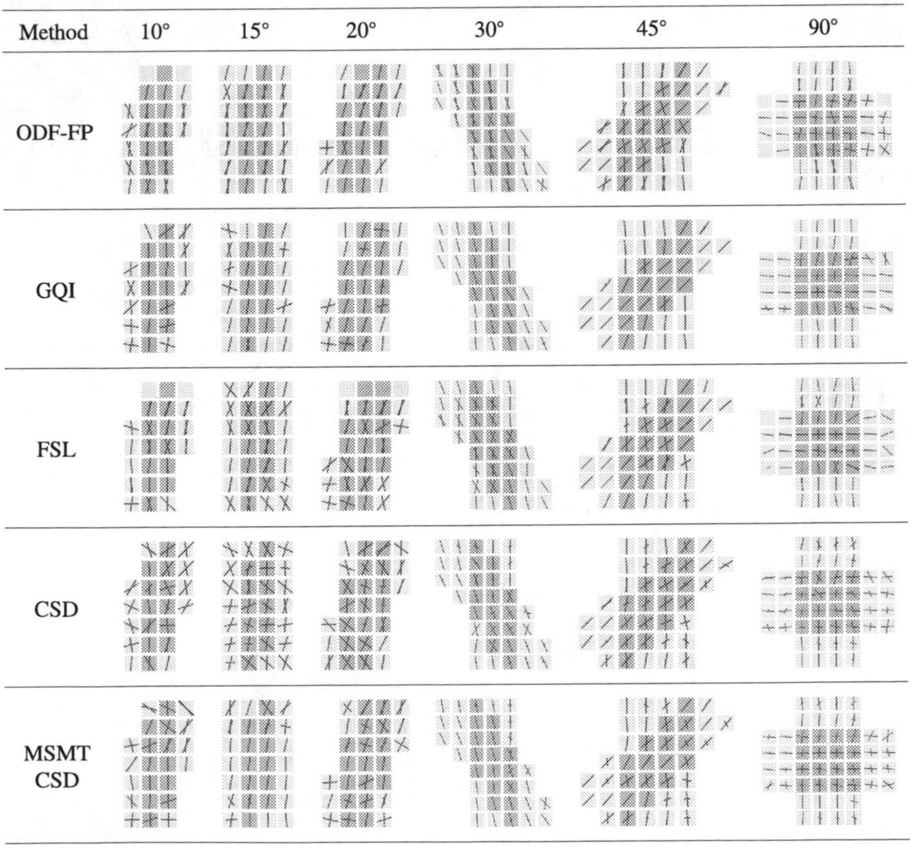

Fig. 2. Samples of fiber directions (red lines) reconstructed with the tested approaches (in rows) for fibers crossing at varied angles (in columns). The regions within the crossing volumes are highlighted in yellow. (Color figure online)

found. We thus obtained percentages of correctly identified voxels with crossing fibers:

$$\text{percentage of crossings found} = \frac{|\text{ voxels with 2 fibers found }|}{|\text{ all voxels in crossing volume }|} \cdot 100\%. \quad (3)$$

For all voxels with correctly identified crossing fibers, we computed maximum divergences $\text{div}_{max} \in [0°, 90°]$ between the estimated fiber directions $u, v \in \mathbb{R}^3$, $\|u\|, \|v\| = 1$ and their respective ground truth directions $u^*, v^* \in \mathbb{R}^3$, $\|u^*\|, \|v^*\| = 1$ as follows:

$$\text{div}_{max}((u, v), (u^*, v^*)) = \max\{\arccos(u \cdot u^*), \arccos(v \cdot v^*)\}, \quad (4)$$

where $u \cdot u^*$ and $v \cdot v^*$ are dot products. Next, we computed the percentages of voxels where the respective maximum divergence did not exceed arbitrary tolerance levels $\varepsilon = 10°, 15°, 20°$. This way we were able to determine fiber reconstruction accuracy regardless of the numbers of reconstructed fibers per voxel:

$$\text{reconstruction accuracy} = \frac{|\text{ voxels with div}_{max} \leq \varepsilon |}{|\text{ all voxels in crossing volume }|} \cdot 100\%. \quad (5)$$

To compare ODF-FP with other approaches, we processed our Diffusion-Weighted Images (DWIs) using a selection of common techniques detailed below:

- **Generalized Q-sampling Imaging** (GQI) [30]—We ran the DSI Studio [31] implementation of the algorithm with the diffusion sampling length ratio $l = 1.2$ (ODF sharpness parameter) adjusted empirically.
- **FSL Bedpostx** [5, 12]—We chose the multi-tensor diffusion model with maximally 2 zeppelins [19] per voxel and we set the sampling mechanism parameters to its default values.
- **Constrained Spherical Deconvolution** (CSD) [25]. We calculated the response function using MRtrix3 [24] implementation of the Tournier algorithm (`dwi2response tournier`) [23] for the highest b-shell ($b = 4000 \text{ s/mm}^2$). We then estimated the fiber orientation distributions with the CSD method (`dwi2fod csd`) [25] and we extracted the directions of maximally 2 highest peaks per voxel (`sh2peaks`).
- **Multi-Shell Multi-Tissue CSD** (MSMT CSD) [25]. As above, with the response functions calculated using the Dhollander algorithm (`dwi2response dhollander`) [8] on all 4 b-shells, and the fiber orientation distributions estimated with the MSMT CSD method (`dwi2fod msmt_csd`) [15] from "white matter"[1] response function only.

---

[1] We put "white matter" in quotation marks to emphasize that our synthetic fibers served as simplified models of WM tissue. Also note that the studied diffusion phantoms did not contain structures representing gray matter or corticospinal fluid, hence the omission of their respective contributions to the diffusion-weighted signal.

28   P. Filipiak et al.

3 Results

Numbers of Fibers Found. All tested methods were able to correctly identify pairs of fibers, even in 10-degree crossings (Fig. 2). However, the mean percentages of crossings found (averaged over all tested ROIs) varied among the reconstruction techniques (Fig. 3) with the visible dominance of ODF-FP (80 ± 8%) over MSMT CSD (64 ± 23%), FSL (63 ± 17%), CSD (60 ± 45%), and GQI (25 ± 14%). Note that the most common error was mistaking a pair of fibers for a single fiber, especially in the GQI-based reconstructions (Figs. 2 and 3).

Fiber Directions Found. ODF-FP reached the highest reconstruction accuracy in fiber crossings at 10, 15, 20, and 30°, while CSD-based approaches scored highest

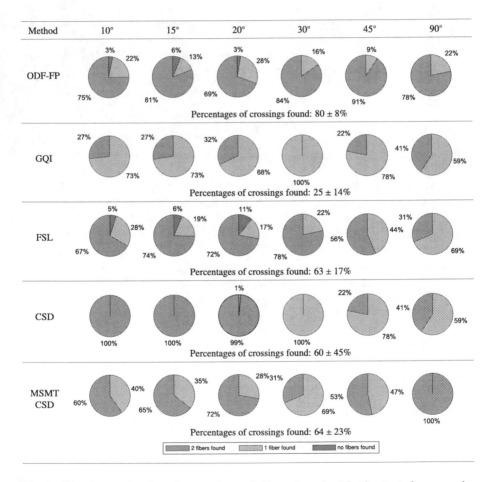

Fig. 3. Pie charts showing the numbers of fibers found with the tested approaches (in rows) for fibers crossing at varied angles (in columns): '2 fibers found' (green) is correct, 1 (yellow) or no fibers (red) are incorrect. Averages and standard deviations of the correctly identified crossings are given for each method. (Color figure online)

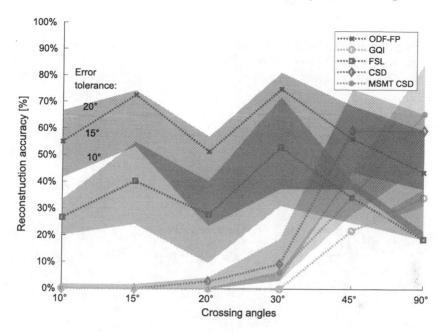

Fig. 4. Reconstruction accuracy in fiber crossings at $10°$, $15°$, $20°$, $30°$, $45°$, and $90°$ estimated for the error tolerance levels $\varepsilon = 10°, 15°, 20°$.

in 45 and 90° (Fig. 4). Assuming the $\varepsilon = 20°$ tolerance for the maximum divergence between the reconstructed fiber directions and the ground truth, ODF-FP maintained the highest mean level of reconstruction accuracy among all the tested angles ($68\pm9\%$), outperforming FSL ($44\pm18\%$), CSD ($28\pm34\%$), MSMT CSD ($22 \pm 35\%$), and GQI ($10 \pm 16\%$).

Interestingly, the pairs of fibers found by GQI, CSD, and MSMT CSD in the challenging 10–20° range (Fig. 3) came with low reconstruction accuracy (Fig. 4) indicating a prevalence of false positives in these approaches.

Distributions of Maximum Divergence. Histograms of maximum divergence provided the most detailed view of the reconstruction accuracy (Figs. 5 and 6). In particular, all pairs of fibers obtained with ODF-FP and FSL, regardless of the crossing angle, had relatively low divergence. The respective medians of div_{max} spanned 9°–15° (ODF-FP) or 9°–20° (FSL). GQI rarely identified crossing fibers, and when it did, the divergence was dispersed between 0° (perfect reconstruction accuracy) and 76° (very low reconstruction accuracy). Both CSD-based approaches reconstructed visibly more crossing fibers than GQI, although the fiber directions were off in all crossing angles below 45°. Nonetheless, MSMT CSD performed best at 90°.

4   Discussion

Breaking the Narrow-Crossing-Angle Barrier. Many current approaches fail to reconstruct fibers crossing at narrow angles below 45° due to limited angu-

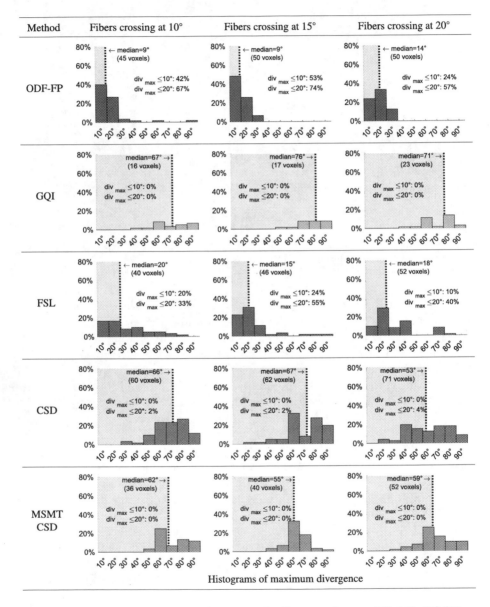

Fig. 5. Distributions of maximum divergence for fiber crossings at 10°, 15°, 20° (Phantom 1). The dotted lines show median values. The numbers of voxels with correctly identified crossing fibers are given in parentheses.

lar resolution and ODF peak width [3,13]. In a typical scenario, narrow-angled crossings are barely distinguishable from a single fiber regardless of the angular precision of diffusion sampling [6,14,22,25], as it was shown in the 3D Validation of Tractography with Experimental MRI (3D-VoTEM) [20] challenge.

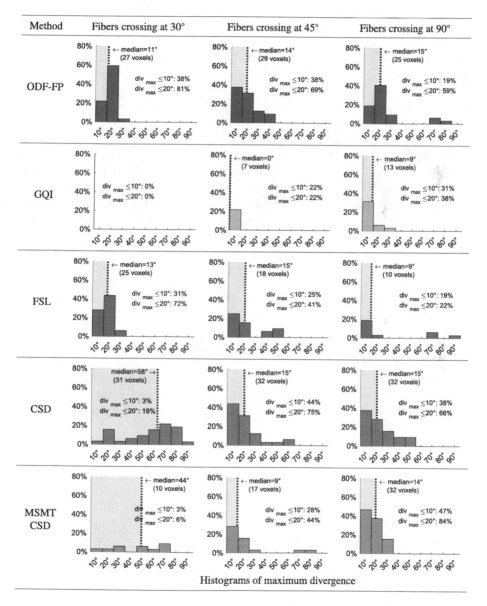

Fig. 6. Distributions of maximum divergence for fiber crossings at 30°, 45°, 90° (Phantom 2). The dotted lines show median values. The numbers of voxels with correctly identified crossing fibers are given in parentheses.

Difficulty in reconstructing fibers crossing at narrow angles hampers visualization of several neural structures including the base of the corpus callosum [7] or retinogeniculate visual pathways [10]. An earlier study showed that ODF-FP was able to break the narrow-crossing-angle barrier on synthetic diffusion data [9]. Our results demonstrate that the same barrier can be broken in modern diffusion phantoms with the textile fibers approaching the anatomical scale of axons. Indeed, ODF-FP successfully identified most pairs of fiber bundles crossing at the angles as narrow as 10°. This observation supports the argument that entire ODF shapes (considered in ODF-FP) contain more information about the underlying fiber structure than merely ODF local maxima.

Pitfalls of Measuring Crossing Angles of Reconstructed Fibers. In this study, we purposely avoided estimation of angles between the reconstructed crossing fibers. Despite their intuitive interpretation, angles between reconstructed fibers are suboptimal measures of performance. In practice, wrongly reconstructed fiber directions may happen to cross at the desired angle, although on a different plane than the ground truth, which would confound the results. Furthermore, it is a common scenario that pairs of fibers crossing at a narrow angle are wrongly identified as a single fiber which prevents estimation of a crossing angle in such cases. Finally, methods that recognize less crossing fibers yet with higher precision would be privileged over other methods that recognize more crossing fibers yet with lower precision. Taking into account the above pitfalls and the ultimate goal of seeking the reconstruction methods that would better support WM fiber tractography, we defined the reconstruction accuracy measure that is free from the biases of the fiber crossing angle.

Limitations. Our study might have underestimated the reconstruction accuracy of CSD-based approaches and GQI, all of which were designed to reach their peak performance in vivo. In particular, we were not able to provide the CSD MSMT algorithm with the full information it expected, since the tested phantoms did not contain structures that would represent real variability of brain tissue, especially in gray matter. As another limitation, we point out that the synthetic fiber bundles in diffusion phantoms did not reflect undulations, fanning, kissing, volume variations, or angular dispersion of axons—frequently observed in brain WM [14,18]—which are known confounding factors for dMRI-based reconstructions that need to be addressed in vivo. Modeling the above examples of higher-level variability in the neural tissue sets an ambitious goal for future phantom designs. Nonetheless, in this work, we were able to quantify the relative impact of crossing angles on overall performance of the tested methods.

Future Work. Our results add to the growing evidence that ODF-FP has potential to visualize WM fibers crossing at challenging narrow angles in vivo. Future work should evaluate the performance of ODF-FP in brain tractography. In addition, the high performance of the CSD-based methods when reconstructing fiber crossings at 45° and 90° (also reported earlier on synthetic data [6]) indicates a potential in combining these techniques with ODF-FP.

5   Conclusion

In this study, we tested the performance of dMRI-based fiber reconstruction using physical diffusion phantoms with varied crossing angles spanning 10–90 degrees. Our results show that ODF-FP is able to correctly identify most of the crossing fibers regardless of the crossing angle and provide stable mean reconstruction accuracy.

Acknowledgements. This project was supported in part by the National Institutes of Health (NIH, R01 EB028774 and R01 NS082436) under the rubric of the Center for Advanced Imaging Innovation and Research (CAI2R, https://www.cai2r.net), a NIBIB Biomedical Technology Resource Center (NIH P41 EB017183).

Source Code Availability Statement. The Python code of ODF-FP implemented as an extension of the DIPY library is available at https://github.com/filipp02/dipy_odffp.

References

1. Baete, S.H., Cloos, M.A., Lin, Y.C., Placantonakis, D.G., Shepherd, T., Boada, F.E.: Fingerprinting orientation distribution functions in diffusion MRI detects smaller crossing angles. Neuroimage **198**, 231–241 (2019)
2. Baete, S.H., Yutzy, S., Boada, F.E.: Radial Q-space sampling for DSI. Magn. Reson. Med. **76**(3), 769–780 (2016)
3. Barnett, A.: Theory of Q-ball imaging redux: implications for fiber tracking. Magn. Reson. Med.: Off. J. Int. Soc. Magn. Reson. Med. **62**(4), 910–923 (2009)
4. Basser, P.J., Pajevic, S., Pierpaoli, C., Duda, J., Aldroubi, A.: In vivo fiber tractography using DT-MRI data. Magn. Res. Med. **44**(4), 625–632 (2000)
5. Behrens, T.E., Berg, H.J., Jbabdi, S., Rushworth, M.F., Woolrich, M.W.: Probabilistic diffusion tractography with multiple fibre orientations: what can we gain? Neuroimage **34**(1), 144–155 (2007)
6. Canales-Rodríguez, E.J., et al.: Sparse wars: a survey and comparative study of spherical deconvolution algorithms for diffusion MRI. Neuroimage **184**, 140–160 (2019)
7. Deslauriers-Gauthier, S., Marziliano, P., Paquette, M., Descoteaux, M.: The application of a new sampling theorem for non-bandlimited signals on the sphere: Improving the recovery of crossing fibers for low b-value acquisitions. Med. Image Anal. **30**, 46–59 (2016)
8. Dhollander, T., Mito, R., Raffelt, D., Connelly, A.: Improved white matter response function estimation for 3-tissue constrained spherical deconvolution. In: Proceedings of the International Society for Magnetic Resonance in Medicine, vol. 555 (2019)
9. Filipiak, P., Shepherd, T., Lin, Y.C., Placantonakis, D.G., Boada, F.E., Baete, S.H.: Performance of orientation distribution function-fingerprinting with a biophysical multicompartment diffusion model. Magn. Reson. Med. **88**(1), 418–435 (2022)
10. He, J., et al.: Comparison of multiple tractography methods for reconstruction of the retinogeniculate visual pathway using diffusion MRI. Hum. Brain Mapp. **42**(12), 3887–3904 (2021)

11. Jelescu, I.O., Budde, M.D.: Design and validation of diffusion MRI models of white matter. Front. Phys. **5**, 61 (2017)
12. Jenkinson, M., Beckmann, C.F., Behrens, T.E., Woolrich, M.W., Smith, S.M.: FSL. Neuroimage **62**(2), 782–790 (2012)
13. Jensen, J.H., Helpern, J.A.: Resolving power for the diffusion orientation distribution function. Magn. Reson. Med. **76**(2), 679–688 (2016)
14. Jeurissen, B., Descoteaux, M., Mori, S., Leemans, A.: Diffusion MRI fiber tractography of the brain. NMR Biomed. **32**(4), e3785 (2019)
15. Jeurissen, B., Tournier, J.D., Dhollander, T., Connelly, A., Sijbers, J.: Multi-tissue constrained spherical deconvolution for improved analysis of multi-shell diffusion MRI data. Neuroimage **103**, 411–426 (2014)
16. Kellner, E., Dhital, B., Kiselev, V.G., Reisert, M.: Gibbs-ringing artifact removal based on local subvoxel-shifts. Magn. Res. Med. **76**(5), 1574–1581 (2016)
17. Maier-Hein, K.H., et al.: The challenge of mapping the human connectome based on diffusion tractography. Nat. Commun. **8**(1), 1–13 (2017)
18. Nilsson, M., Lätt, J., Ståhlberg, F., van Westen, D., Hagslätt, H.: The importance of axonal undulation in diffusion MR measurements: a Monte Carlo simulation study. NMR Biomed. **25**(5), 795–805 (2012)
19. Panagiotaki, E., Schneider, T., Siow, B., Hall, M.G., Lythgoe, M.F., Alexander, D.C.: Compartment models of the diffusion MR signal in brain white matter: a taxonomy and comparison. Neuroimage **59**(3), 2241–2254 (2012)
20. Schilling, K.G., et al.: Limits to anatomical accuracy of diffusion tractography using modern approaches. Neuroimage **185**, 1–11 (2019)
21. Schneider, W., Pathak, S., Wu, Y., Busch, D., Dzikiy, D.J.: Taxon anisotropic phantom delivering human scale parametrically controlled diffusion compartments to advance cross laboratory research and calibration. In: ISMRM 2019 (2019)
22. Tournier, J.D., Calamante, F., Connelly, A.: MRtrix: diffusion tractography in crossing fiber regions. Imaging Syst. Technol. **22**(1), 53–66 (2012)
23. Tournier, J.D., Calamante, F., Connelly, A.: Determination of the appropriate b value and number of gradient directions for high-angular-resolution diffusion-weighted imaging. NMR Biomed. **26**(12), 1775–1786 (2013)
24. Tournier, J.D., Smith, R., Raffelt, D., Tabbara, R., Dhollander, T., et al.: MRtrix3: a fast, flexible and open software framework for medical image processing and visualisation. Neuroimage **202**, 116137 (2019)
25. Tournier, J.D., Yeh, C.H., Calamante, F., Cho, K.H., Connelly, A., Lin, C.P.: Resolving crossing fibres using constrained spherical deconvolution: validation using diffusion-weighted imaging phantom data. Neuroimage **42**(2), 617–625 (2008)
26. Tuch, D.S., Reese, T.G., Wiegell, M.R., Wedeen, V.J.: Diffusion MRI of complex neural architecture. Neuron **40**(5), 885–895 (2003)
27. Tustison, N.J., Avants, B.B., Cook, P.A., Zheng, Y., et al.: N4ITK: improved N3 bias correction. IEEE Trans. Med. Imaging **29**(6), 1310–1320 (2010)
28. Veraart, J., Novikov, D.S., Christiaens, D., Ades-Aron, B., et al.: Denoising of diffusion MRI using random matrix theory. Neuroimage **142**, 394–406 (2016)
29. Wedeen, V.J., et al.: Diffusion spectrum magnetic resonance imaging (DSI) tractography of crossing fibers. Neuroimage **41**(4), 1267–1277 (2008)
30. Yeh, F.C., Wedeen, V.J., Tseng, W.Y.I.: Generalized Q-sampling imaging. IEEE Trans. Med. Imaging **29**(9), 1626–1635 (2010)
31. Yeh, F.: Diffusion MRI reconstruction in DSI Studio. Advanced Biomedical MRI Lab, National Taiwan University Hospital (2017). http://dsi-studio.labsolver.org/Manual/Reconstruction

Improving Multi-Tensor Fitting with Global Information from Track Orientation Density Imaging

Erick Hernandez-Gutierrez[1]($\boxtimes$) (ID), Ricardo Coronado-Leija[2] (ID),
Alonso Ramirez-Manzanares[3] (ID), Muhamed Barakovic[5] (ID), Stefano Magon[5] (ID),
and Maxime Descoteaux[1,4] (ID)

[1] University of Sherbrooke, Sherbrooke, QC J1K2R1, Canada
erick.hernandez.gutierrez@usherbrooke.ca
[2] New York University School of Medicine, New York, NY 10016, USA
[3] Center of Research in Mathematics, 36023 Guanajuato, GT, Mexico
[4] Imeka Solutions Inc., Sherbrooke, QC J1H 4A7, Canada
[5] Pharma Research and Early Development, Neuroscience and Rare Diseases Roche
Innovation Center Basel, F. Hoffmann-La Roche Ltd., 4070 Basel, CH, Switzerland

Abstract. The number of fascicles per-voxel N is a crucial parameter for multiple-fiber reconstruction methods from diffusion magnetic resonance imaging (dMRI) data, such as the multi-tensor model (MTM). This parameter is especially important when the goal is to provide bundle-specific tissue metrics. However, for MTM, statistical selection methods, such as the F-test, the Akaike and the Bayesian information criteria, tend to overestimate the number of tensors per-voxel for typical multi-shell dMRI acquisitions. In this work, we explore the reasons for this overestimation and propose to combine track orientation density imaging (TODI) with a robust MTM fitting framework to improve the estimation of N per-voxel. We conducted experiments on both synthetic and clinical *in vivo* dMRI data to validate our method, and we observed that tractography seems to be a useful spatial regularizer for N. Our results demonstrate the effectiveness of incorporating TODI for a more accurate and robust estimation of the intra-voxel number of fascicles.

Keywords: Tractography · Multi-Tensor · Regularization

1 Introduction

Similar to the definition of pixel and voxel, *fixel* is the smallest discrete component of a fiber element [1]. Each fixel is parameterized by the voxel in which it resides. Several methods have been developed to estimate properties (directions and tissue metrics) of the local fixels in brain white matter (WM) using diffusion magnetic resonance imaging (dMRI). One of the most commonly used methods is Diffusion Tensor Imaging (DTI) [18], which can be obtained from fast and single b-value acquisitions. However, DTI can only estimate one fixel per-voxel.

M. Karaman et al. (Eds.): CDMRI 2023, LNCS 14328, pp. 35–46, 2023.
https://doi.org/10.1007/978-3-031-47292-3_4

This is insufficient, as WM tissue has a complex structure containing many fiber crossings regions [3,13,23].

More complex methods were developed for a better estimation of the intra-voxel fixels such as the Constrained Spherical Deconvolution (CSD) [14,25,26]. Unfortunately, the main limitation of CSD-based methods is that the diffusion profile (response function) is kept fixed across the WM. Moreover, in contrast with the metrics provided by DTI, fixel metrics provided by CSD lack a straight-forward biological interpretation. Even advanced methods, such as NODDI [31], that consider dispersion, intra- and extra-axonal compartments, only account for a single fiber population. One of the main challenges for the intra-voxel reconstruction methods is to find a balance between complexity and accuracy.

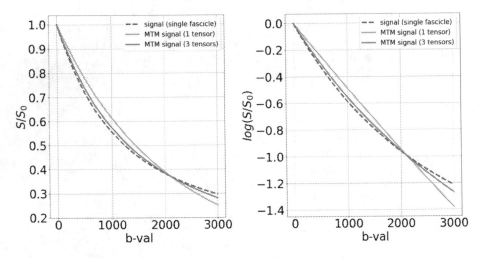

Fig. 1. Comparison between a simulated multi-compartment dMRI signal (red dashed) in a single voxel containing one fascicle, a fitted MRDS with 1 tensor (yellow) and 3 tensors (green). Signals are shown as a function of the b-value. It is desired $N = 1$, but $N = 3$ fits better the signal. Thus, during the model selector in MRDS $N = 3$ is selected instead of $N = 1$ resulting in an overestimation of the number of fascicles. (Color figure online)

The Multi-Tensor Model (MTM) [12], which is an extension of DTI, represents the dMRI signal at each voxel as a discrete mixture of Gaussian diffusion distributions. Unlike DTI, the presence of N diffusion tensors at each voxel involves solving a computationally complex nonlinear problem. Several methods have been proposed to estimate MTM parameters [4,11,15,16,19,21,28]. These works reported that MTM fitting is numerically challenging and unstable. Among these methods, the Multi-Resolution Discrete Search (MRDS) framework [4] presents a viable trade-off between complexity and accuracy when working with short-acquisition-time clinical-like multi-shell dMRI data sets. MRDS has shown to be stable, accurate, and robust-to-noise for the estimation of intra-voxel fixel orientations and diffusion properties [4,20,29].

MRDS is a robust estimator of per-fixel tissue properties that performs better when the number of fixels (N) is known. However, for typical multi-shell dMRI acquisitions, this method struggles to accurately estimate the N that best represents the dMRI signal in the voxel. Figure 1 describes why MRDS, and in general any MTM framework, overestimates N and why this is a hard-to-solve problem. Shortly, for typical multi-shell dMRI acquisitions, MTM estimators tend to overestimate the per-voxel number of fixels because the diffusion tensor does not correctly represent the single fascicle dMRI signal for high b-values ($b > 1000$ s/mm^2). Figure 1 shows a dMRI signal generated using a single fascicle stick-zeppelin-ball multi-compartment model with Watson dispersion [17,31]. It is observed, in logarithm scale, that the dMRI signal decay is non-linear while the tensor signal decay is linear. By including all b-values in the fitting, MRDS increases the number of tensors to reduce the signal reconstruction error, resulting in a solution with $N = 3$ tensors instead of the expected solution with $N = 1$ tensor, because the signal was generated for a single fascicle. Discarding the high b-values of the dMRI signal increases the estimation accuracy for N. However, multi-shell dMRI data is mandatory for the well-posedness of MTM fitting [22], then it is important in general to keep all information, as many clinical and public data sets rely on two or three shells, containing b-values higher than 1000 s/mm^2.

Diffusion tractography is a valuable technique that provides insights into the connections and pathways of axons in the WM [30]. It has found many wide-ranging applications in the study of WM tracts. For example, tractography can be used to regularize parameter estimation, by incorporating information not just from the local measurements in a given voxel, but also from the voxels belonging to the same track. Furthermore, Track Orientation Density Imaging (TODI) is a technique that reconstruct a complete description of the track orientation distribution function (tODF) at each voxel. TODI aims to estimate the tODF throughout the entire brain and can be applied to any tractogram resulting from a fiber tracking algorithm.

Since MRDS already provides robust solutions for $N = 1$, $N = 2$, and $N = 3$ at each voxel. We hypothesize that TODI can play a crucial role by incorporating tractography regularization into the MTM selection of N, as it captures the tracking information at each voxel into a tODF, from which a better estimation of the intra-voxel number of fascicles can be obtained. Using this better N could help MRDS to generate more accurate, robust, and reproducible multi-tensor fields (MTF). In this work, we propose to re-inform MRDS with TODI, improving the overall accuracy when estimating the number of intra-voxel fascicles. Results and implications are presented for synthetic and *in vivo* experiments.

2   Materials and Methods

2.1   Pipeline

Our pipeline consists of five steps for processing dMRI data. 1) Data is denoised, distortion corrected, and up-sampled [9,24]. 2) MRDS is used to estimate the orientations and tensor diffusivities per-fixel from the dMRI data. MRDS provides MTM solutions for $N = 1, 2, 3$ at each voxel. Then, it uses a statistical model selector (AIC, BIC, F-test) to choose the best N. We employed the dmri-explorer [7] to visualize the estimated MTF. 3) The MTF is represented as ODFs [2]. 4) With these ODFs, the iFOD2 [27] tractography algorithm is performed to get a MRDS-based tractogram. In this work, we used the raw tractogram for all the experiments. However, a tractogram filtering method, such as COMMIT [6], could be used to remove potential false positive streamlines at this step of the pipeline. 5) TODI is used to reconstruct the tODF at each voxel and generate a **N**umber-of-**F**iber-**O**rientations (NuFO) map from the tODFs. This map is later used to select the best MRDS solution ($N = 1$, $N = 2$ or $N = 3$) at each voxel.

The TODI tODF is the probability distribution of the streamlines orientations within a voxel. When false positives streamlines are presented in the tractogram, they introduce spurious peaks in the tODF. However, theses peaks are usually small as the false positives streamlines come from overestimated tensors lacking spatial support. Thus, they can be removed by a threshold in the amplitudes of the tODF when computing the NuFO map. Consequently, the resulting TODI NuFO map is a better estimation than the NuFO obtained with AIC, BIC or F-test. Therefore, this TODI NuFO map can be used to improve the model selector step in MRDS. Although this process can become iterative by computing new MRDS-ODFs, MRDS-tractograms, and MRDS-tODFs, from each iteration of MRDS MTFs, in this work we used only a single iteration to demonstrate the potential of this strategy.

In this study, we computed TODI from two different tractograms: one tractogram generated using the particle filtering technique from the fODFs estimated with CSD, and another tractogram generated using the ODFs computed from the MRDS MTF and the iFOD2 algorithm with cutoff = 0.05 (Fig. 2).

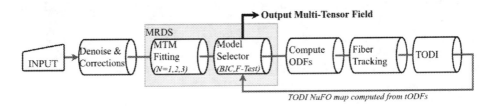

Fig. 2. Diagram of the pipeline used to process the data. It includes the novel model selector informed from TODI.

2.2 Data

Synthetic: To validate the proposed method in controlled scenarios, we used the digital phantom from the ISBI 2012 - HARDI Reconstruction Workshop [5]. This phantom has a structure field with specific arrangements of synthetic fibers that mimic challenging crossing configurations. The structure of the phantom consists of $16{\times}16{\times}5$ voxels and 3 fiber bundles with a 3D configuration of tracts. The dMRI synthetic data was numerically generated in every voxel using the stick-zeppelin-ball model and a Watson distribution with $\kappa = 16$. [17,31]. The acquisition protocol includes 7 $b = 0$ images and 16 b-values with a step-size of 200 between b-value shells arranges as following. There are 10 directions from $b = 100$ to $b = 900$, 30 directions from $b = 1000$ to $b = 1800$ and 60 directions from $b = 2000$ to $b = 3000$, resulting in a total of 567 directions. Here, all bundles share the same ground-truth (GT) values, where for the compartment sizes are $f_{\mathrm{a}} = 0.7$, $f_{\mathrm{e}} = 0.25$, and $f_{\mathrm{w}} = 0.05$ for the intra-axonal, extra-axonal, and free water compartments, respectively, while their GT diffusivities are $D_{\mathrm{a}}^{\|} = D_{\mathrm{e}}^{\|} = 2 \cdot 10^{-3}$ mm^2/s, $D_e^{\perp} = 0.5 \cdot 10^{-3}$ mm^2/s and $D_{\mathrm{w}} = 3 \cdot 10^{-3}$ mm^2/s.

Three TODI-based NuFO maps were generated using tODFs computed from CSD, MRDS and GT tracking. CSD tracking was performed using the fODFs estimated from the CSD method. MRDS tracking was performed using the ODFs obtained from the MRDS MTF. GT tracking refers to perform tractography with iFOD2 in each separated bundle using fODFs from the GT synthetic signal without noise. For GT tracking seeds were homogeneously distributed in a way that each per-bundle tractogram contains around $40,000$ streamlines. Then, we merged the individual per-bundle tractograms to get a full tractogram for the whole phantom. This represents the best possible scenario for the tractography step in the proposed pipeline, setting an upper-bound in the performance of the proposed method (Fig. 3).

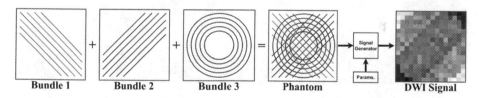

| Bundle 1 | Bundle 2 | Bundle 3 | Phantom | | DWI Signal |

Fig. 3. Illustration of the structure of the phantom and its bundles. The dMRI signal includes noise and dispersion. All bundles have the same diffusivities.

In vivo: The data set of this study was acquired using a 3T MRI Philips Healthcare System. Twenty healthy control (HC) subjects were scanned using a multishell diffusion-weighted protocol with 3 non-zero b-shells ($b = 300$, 1000 and 2000 mm^2/s with 8, 32, and 60 directions, respectively), 7 $b = 0$ images, and 2 mm isotropic resolution [8].

Two experiments were designed to test the stability and reproducibility of the proposed pipeline. The first experiment involves applying our pipeline to the dMRI of a single subject scanned 6 times with a 4-week interval ($\pm$ 1 week) over 6 months. The second experiment involves applying our pipeline to the dMRI of 17 distinct subjects, each scanned once. For the evaluation of the two experiments, we employed the following metrics:

- **Mean**: Refers to the mean NuFO map of the 6 scans for the single subject, and the mean NuFO map of the 17 subjects in the HC group scanned once. These NuFOs are continuous intensity maps.
- **Rounded Mean**: This is the rounded value to the closest integer of the mean NuFO map, which offers an integer-valued NuFO map that can be used in MRDS.
- **Success Rate (SR)**: Indicates the proportion of times, per-voxel, in which the rounded NuFO map of for a single scan is equal to rounded mean NuFO. SR is used to measure of the variability across the different scans for the single subject, or the variability between different subjects for the case of the HC group.

Following previous studies [3,14,23], we investigated the proportions of voxels with $N = 1$, $N = 2$, and $N = 3$ fascicles, selected by all the different strategies explored in this work. We report averages of these proportions in WM for all the scans in the single subject and for all the subjects in the HC group. We also report averages for the success rate in each case.

3   Results

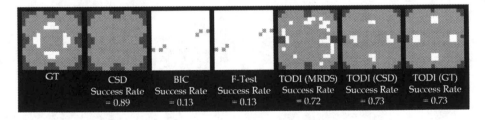

Fig. 4. Comparison of the maps showing the estimated number of fascicles obtained with different methods on synthetic data. The success rate with respect to the GT is included.

In Fig. 4 we compare the NuFO maps estimated with different methods on the synthetic data. The effectiveness of each method to recover the structural information of the phantom is shown visually and quantitatively. Statistical model selector methods BIC and F-test over-estimate the number of tensors N, choosing $N = 3$ in almost all the voxels. CSD does not recover $N = 3$ for any voxel, clearly underestimating the number of fascicles. Regarding TODI-based methods, the NuFO maps show that TODI (CSD) performance is comparable with TODI (GT), while TODI (MRDS) has difficulties recovering accurately regions with $N = 3$. Nonetheless, TODI (MRDS) is quantitatively similar to TODI (CSD) and TODI (GT), with significant better results than statistical model selector criteria such as BIC and F-test. The main difference between TODI

(CSD) and TODI (MRDS) is qualitative. The lower success rate for TODI-based methods compared with CSD can be explained by the sub-optimality of the phantom for tractography, lacking anatomical features used by tractography methods to control the behavior of streamlines in voxels at edges. For this reason NuFO maps from tractography based methods have dilated edges, reducing their success rate.

Figure 5 illustrates a grid of maps for the 3 different measures described before obtained from the *in vivo* experiments using 5 different model selectors. We included the two statistical methods already proposed in MRDS: BIC and F-test. Besides, we incorporated another NuFO map obtained with CSD. The images in Fig. 5 highlight the performance and variability of each method. In particular, Fig. 5 shows that TODI (MRDS) performs similar as TODI (CSD), however, the NuFO map obtained with TODI (MRDS) seems noisier and than the map obtained with TODI (CSD).

Table 1 summarizes the results for the estimations of the number of fascicles per voxel in the single subject and the group of subjects. CSD shows a large proportion of voxels with a single fiber, while showing almost null proportion of $N = 3$ fiber crossings. BIC show a large proportion of $N = 3$ fiber crossings, and similarly smaller proportions for single fiber and $N = 2$ fiber crossings. F-test show similar proportions as BIC, but with a more balanced proportions for $N = 2$ and $N = 3$ fiber crossings. Lastly, TODI (CSD) and TODI (MRDS) show similar results in proportions and success rate values. In general, all methods reveal consistent proportion values between the experiments involving the single HC subject scanned 6 times and the group of subjects scanned once, with difference no larger than 10%. On the other hand, the success rate is not as consistent, we observed significant lower SR values for the group than for the single subject. Although this could indicate a larger biological variability, it could also be the result of larger registration errors for the population than for the single subject scanned 6 times.

Table 1. Average of the proportions (prop. %) and success rate (SR) for different model selectors. Both proportions and success rate are reported for the experiment with the single subject (sub.) and the HC group (grp.).

Method	$N = 1$ prop. (%) sub. | grp	SR sub. | grp	$N = 2$ prop. (%) sub. | grp	SR sub. | grp	$N = 3$ prop. (%) sub. | grp	SR sub. | grp
CSD	60.3 | 68.9	0.87 | 0.68	17.3 | 28.1	0.61 | 0.36	0.96 | 2.92	0.19 | 0.04
BIC	12.1 | 14.6	0.52 | 0.47	18.2 | 22.3	0.52 | 0.33	56.2 | 63.1	0.88 | 0.50
F-test	12.5 | 14.7	0.52 | 0.50	39.4 | 44.5	0.72 | 0.46	34.4 | 40.7	0.68 | 0.34
TODI (CSD)	20.5 | 22.8	0.61 | 0.57	37.4 | 42.1	0.76 | 0.52	27.7 | 35.0	0.66 | 0.32
TODI (MRDS)	28.7 | 33.5	0.60 | 0.56	40.7 | 47.1	0.79 | 0.50	16.1 | 19.3	0.43 | 0.15

Figure 6 shows the MTF for the different methods to select N. Figure 6 highlights a common ROI from the tensor fields to qualitatively compare their spatial

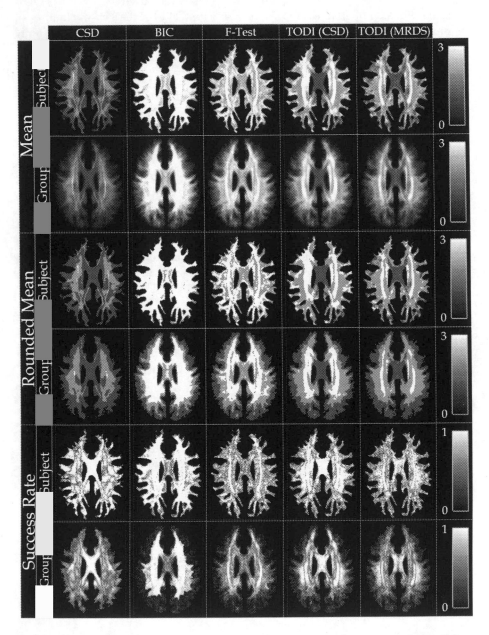

Fig. 5. Mean, rounded mean and success rate (top to bottom) measures obtained from a single subject scanned 6 times and a group of 17 subjects of the *in vivo* data set using different NuFO maps: BIC, F-test, CSD, TODI-CSD and TODI-MRDS (left to right).

coherence. This zoomed-in section provides a detailed view of the complex distribution of tensors in a particular brain region with several fiber crossings. Here, TODI (CSD) and TODI (MRDS) tensors exhibit a significant level of spatial coherence throughout the voxels containing $N = 2$ and $N = 3$ fiber fascicles. On the other hand, BIC shows inconsistencies in some voxels with $N = 3$ tensors where it is clear that several of these tensors do not have support from the neighborhood. For this particular ROI, F-test shows a similar but noisier result compared to TODI (CSD) and TODI (MRDS). CSD seems to underestimate the number of fascicles in comparison to the other methods, as regions with $N = 3$ are highly reduced, with several spatially supported tensors in the y-axis (green) missing.

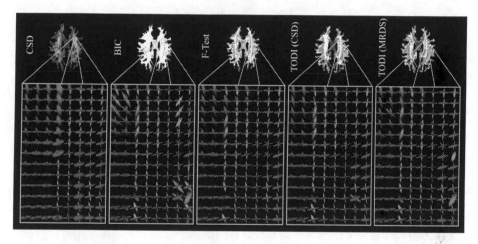

Fig. 6. A comparison of different NuFO maps is shown in a ROI where 2- and 3-fiber crossings are present. The MTFs for $N = 1$, $N = 2$ and $N = 3$ were fitted with MRDS and visualized with the dmri-explorer [7]. Tensors are colored according to the principal diffusion direction of the tensor.

4   Discussion

An accurate and robust NuFO map is important when using the MTM fixel-based metrics as input for a tractometry analysis. [10]. Our experiments for the phantom dMRI data set showed that the proposed method is effective to accurately estimate the number of fascicles per-voxel at challenging crossing configurations. However, it is important to mention that the phantom is an idealized scenario lacking realistic anatomical structures and where a GT tractogram could be computed.

Figure 5 shows that BIC overestimates the number of fascicles per-voxel, with Table 1 indicating a combined proportions of voxels with $N = 2$ and $N = 3$

around 74%. F-test presents a less prominent, but still overestimating behavior. CSD is clearly underestimating the number of fascicles as Fig. 5 and Table 1 show almost no $N = 3$ voxels. On the other hand, TODI (CSD) estimates the combined proportion for $N = 2$ and $N = 3$ voxels around 64%, which is congruent with previous studies [3,13,23], while TODI (MRDS) shows comparable results. For Table 1, the sum of the rows of each method may not be equal to 100% because there are voxels in which methods provided $N = 0$ as solution, the remaining proportion corresponds to these voxels.

TODI (CSD) and TODI (MRDS) provide similar results in Figs 4 and 5 for synthetic and real data, however TODI (MRDS) NuFO maps are noisier. Two factors could be influencing these results. First, the MRDS-based tractogram is computed using BIC as preliminary model selector, which overestimates the number of fascicles, generating a sub-optimal tractogram with considerable false positive streamlines. Second, the MRDS-based tractogram is computed using the iFOD2 algorithm, which is not tuned to work with the ODFs derived from multi-tensors, but it is tuned for fODFs derived from CSD. Because the lack of GT for the real dMRI data and the small differences between TODI (CSD) and TODI (MRDS), it is difficult to determine which method is better. However, it is important to notice that TODI (CSD) requires additional parameters and steps in the pipeline such as estimating a response function and fitting CSD to the data. For TODI (MRDS), these additional steps are omitted because the fiber tracking is performed using the multi-tensor ODFs, resulting in less required parameters and less processing time (no extra steps).

Performing fiber tracking from the MRDS ODFs using the iFOD2 method seems promising, however iFOD2 was developed for CSD fODFs and its better suited for them. Then, it is imperative to explore other tracking techniques designed for MTFs. The tractograms used in our study were not pruned from potential false positive streamlines. Therefore, experiments are needed to test methods for tractogram filtering, such as COMMIT. Future work will focus in using our pipeline on subjects with neurological diseases, such as multiple sclerosis, for a better characterization of bundle-specific tissue metrics.

5   Conclusion

Using tractography to regularize the estimation of the number of fixels at every voxel during MTM fitting improves spatial consistency, reproducibility, and robustness. By combining TODI and MRDS, a general improvement in the accuracy of the NuFO maps is achieved in comparison to statistical selector methods, such as BIC and F-test, reducing overfitting. Finally, we observed that using the MTF estimated by MRDS as the local microstructure model for fiber tracking shows promising results to create a complete pipeline to re-inform the model selector in MRDS.

References

1. https://mrtrix.readthedocs.io/en/dev/concepts/fixels_dixels.html
2. http://hardi.epfl.ch/static/events/2012_ISBI/data_format.html
3. Behrens, T., Berg, H.J., Jbabdi, S., Rushworth, M., Woolrich, M.: Probabilistic diffusion tractography with multiple Fibre orientations: what can we gain? Neuroimage **34**(1), 144–155 (2007)
4. Coronado-Leija, R., Ramirez-Manzanares, A., Marroquin, J.L.: Estimation of individual axon bundle properties by a multi-resolution discrete-search method. Med. Image Anal. **42**, 26–43 (2017)
5. Daducci, A., et al.: Quantitative comparison of reconstruction methods for intravoxel fiber recovery from diffusion MRI. IEEE Trans. Med. Imaging **33**(2), 384–399 (2014)
6. Daducci, A., Palu, A.D., Lemkaddem, A., Thiran, J.P.: COMMIT: convex optimization modeling for microstructure informed tractography. IEEE Trans. Med. Imaging **34**(1), 246–257 (2015)
7. E., H.G., C., P., R., C.L., M., D.: Real-time rendering of massive multi-tensor fields using modern OpenGL. In: Proceedings of the International Society for Magnetic Resonance in Medicine (2022)
8. Edde, M., et al.: High-frequency longitudinal white matter diffusion- and myelin-based MRI database: Reliability and variability. Hum. Brain Mapp. **44**(9), 3758–3780 (2023)
9. Gudbjartsson, H., Patz, S.: The Rician distribution of noisy MRI data. Magn. Reson. Med. **34**(6), 910–914 (1995)
10. Hernandez-Gutierrez, E., et al.: Multi-tensor fixel-based metrics in tractometry: application to multiple sclerosis. In: Proceedings of the 2023 Annual Meeting of the Organization for Human Brain Mapping. Montreal, Canada (2023)
11. Hosey, T., Williams, G., Ansorge, R.: Inference of multiple fiber orientations in high angular resolution diffusion imaging. Magn. Reson. Med. **54**(6), 1480–1489 (2005)
12. Inglis, B., Bossart, E., Buckley, D., Wirth, E., Mareci, T.: Visualization of neural tissue water compartments using biexponential diffusion tensor MRI. Magn. Reson. Med. **45**(4), 580–587 (2001)
13. Jeurissen, B., Leemans, A., Tournier, J.D., Jones, D.K., Sijbers, J.: Investigating the prevalence of complex fiber configurations in white matter tissue with diffusion magnetic resonance imaging. Hum. Brain Mapp. **34**(11), 2747–2766 (2012)
14. Jeurissen, B., Tournier, J.D., Dhollander, T., Connelly, A., Sijbers, J.: Multi-tissue constrained spherical deconvolution for improved analysis of multi-shell diffusion MRI data. Neuroimage **103**, 411–426 (2014)
15. Kreher, B.W., Schneider, J.F., Mader, I., Martin, E., Hennig, J., Il'yasov, K.A.: Multitensor approach for analysis and tracking of complex fiber configurations. Magn. Reson. Med. **54**(5), 1216–1225 (2005)
16. Melie-García, L., Canales-Rodríguez, E.J., Alemán-Gómez, Y., Lin, C.P., Iturria-Medina, Y., Valdés-Hernández, P.A.: A bayesian framework to identify principal intravoxel diffusion profiles based on diffusion-weighted MR imaging. Neuroimage **42**(2), 750–770 (2008)
17. Panagiotaki, E., Schneider, T., Siow, B., Hall, M.G., Lythgoe, M.F., Alexander, D.C.: Compartment models of the diffusion MR signal in brain white matter: a taxonomy and comparison. Neuroimage **59**(3), 2241–2254 (2012)

18. Basser, P.J., Mattiello, J., LeBihan, D.: MR diffusion tensor spectroscopy and imaging. Biophys. J . **66**(1), 259–267 (1994)
19. Ramirez-Manzanares, A., Rivera, M., Vemuri, B., Carney, P., Mareci, T.: Diffusion basis functions decomposition for estimating white matter intravoxel fiber geometry. IEEE Trans. Med. Imaging **26**(8), 1091–1102 (2007)
20. Rojas-Vite, G., et al.: Histological validation of per-bundle water diffusion metrics within a region of fiber crossing following axonal degeneration. Neuroimage **201**, 116013 (2019)
21. Scherrer, B., et al.: Characterizing the distribution of anisotropic micro-structural environments with diffusion-weighted imaging (DIAMOND). In: Mori, K., Sakuma, I., Sato, Y., Barillot, C., Navab, N. (eds.) MICCAI 2013. LNCS, vol. 8151, pp. 518–526. Springer, Heidelberg (2013). https://doi.org/10.1007/978-3-642-40760-4_65
22. Scherrer, B., Warfield, S.K.: Parametric representation of multiple white matter fascicles from cube and sphere diffusion MRI. PLoS ONE **7**(11), e48232 (2012)
23. Schilling, K., Gao, Y., Janve, V., Stepniewska, I., Landman, B.A., Anderson, A.W.: Can increased spatial resolution solve the crossing fiber problem for diffusion MRI? NMR Biomed. **30**(12), e3787 (2017)
24. Theaud, G., Houde, J.C., Boré, A., Rheault, F., Morency, F., Descoteaux, M.: TractoFlow: a robust, efficient and reproducible diffusion MRI pipeline leveraging nextflow & amp singularity. Neuroimage **218**, 116889 (2020)
25. Tournier, J.D., Calamante, F., Connelly, A.: Robust determination of the Fibre orientation distribution in diffusion MRI: non-negativity constrained super-resolved spherical deconvolution. Neuroimage **35**(4), 1459–1472 (2007)
26. Tournier, J.D., Calamante, F., Gadian, D.G., Connelly, A.: Direct estimation of the fiber orientation density function from diffusion-weighted MRI data using spherical deconvolution. Neuroimage **23**(3), 1176–1185 (2004)
27. Tournier, J.D., Calamante, F., Connelly, A.: Improved probabilistic streamlines tractography by 2nd order integration over fibre orientation distributions. Proc. Intl. Soc. Mag. Reson. Med. (ISMRM) **18** (2010)
28. Tuch, D.S., Reese, T.G., Wiegell, M.R., Makris, N., Belliveau, J.W., Wedeen, V.J.: High angular resolution diffusion imaging reveals intravoxel white matter fiber heterogeneity. Magn. Reson. Med. **48**(4), 577–582 (2002)
29. Villaseñor, P.J., et al.: Multi-tensor diffusion abnormalities of gray matter in an animal model of cortical dysplasia. Front. Neurol. **14**, 1124282 (2023)
30. Zhang, F., et al.: Quantitative mapping of the brain's structural connectivity using diffusion MRI tractography: a review. Neuroimage **249**, 118870 (2022)
31. Zhang, H., Schneider, T., Wheeler-Kingshott, C.A., Alexander, D.C.: NODDI: practical in vivo neurite orientation dispersion and density imaging of the human brain. Neuroimage **61**(4), 1000–1016 (2012)

BundleSeg: A Versatile, Reliable and Reproducible Approach to White Matter Bundle Segmentation

Etienne St-Onge[1]([✉]), Kurt G Schilling[2,3], and Francois Rheault[4]

[1] Department of Computer Science and Engineering, Université du Québec en Outaouais, Saint-Jérôme, Québec, Canada
Etienne.st-onge@uqo.ca
[2] Department of Radiology and Radiological Sciences, Vanderbilt University Medical Center, Nashville, TN, USA
[3] Vanderbilt University Institute of Imaging Science, Vanderbilt University, Nashville, TN, USA
[4] Medical Imaging and Neuroinformatic (MINi) lab, Université de Sherbrooke, Sherbrooke, Québec, Canada

Abstract. This work presents BundleSeg, a reliable, reproducible, and fast method for extracting white matter pathways. The proposed method combines an iterative registration procedure to a recently developed precise streamline search algorithm that enables efficient segmentation of streamlines without the need for tractogram clustering or simplifying assumptions. We show that BundleSeg achieves improved repeatability and reproducibility than state-of-the-art segmentation methods, with significant speed improvements. The enhanced precision and reduced variability in extracting white matter connections offer a valuable tool for neuroinformatic studies, increasing the sensitivity and specificity of tractography-based studies of white matter pathways.

Keywords: Tractography · White matter bundle · Segmentation · Registration

1 Introduction

Accurate segmentation of tractography bundles is crucial for advancing our understanding of brain structure and organization. Grouping streamlines into white matter (WM) bundles is a common practice in leveraging analysis from Diffusion-Weighted MRI (DW-MRI) models. This allows for targeted WM analysis and cohort comparison along known brain connections. Improving the reliability and reproducibility of bundle segmentation methods can effectively reduce variability in longitudinal and group analyses, thereby enhancing the overall statistical power.

Several methods have been developed to virtually dissect WM bundles from a full brain tractogram. One of the earliest approaches involves the use of region-of-interest (ROI) to virtually dissect tractography results from specific anatomical

M. Karaman et al. (Eds.): CDMRI 2023, LNCS 14328, pp. 47–57, 2023.
https://doi.org/10.1007/978-3-031-47292-3_5

locations [3,12,22,23,26]. While these methods provide some control over the segmentation process, it is often limited by the subjectivity of manually drawn ROIs, or atlas resolution, and may not capture the complexity and geometry of WM pathways [15,16]. Another approach involves clustering algorithms that group streamlines based on similarity to an existing streamlines template, typically using metrics such as spatial proximity, shape, or fiber orientation [8,11,13]. However, most of these approaches are sensitive to initialization and may yield inconsistent results due to variability in the choice of parameters or seed points.

Additionally, machine learning-based techniques have been recently explored, where models are trained on labeled datasets to predict the presence of specific WM bundles [2,24,25]. Although these methods have shown promising results, they heavily rely on the availability of high-quality pre-segmented data and may be limited toward generalization to different populations or imaging protocols. Therefore, there is a need for an improved method that can address these limitations and provide a more reliable and reproducible tractography bundle.

To summarize, current techniques are challenged by limited repeatability (i.e. inconsistent results when run twice on the same dataset), reproducibility (i.e. differences in scan-rescan or longitudinal acquisitions), typically due to time constraints in manual bundle segmentation or in automated segmentation methods. However, repeatable, reproducible, and anatomically accurate approaches are essential in both research and clinical tractography applications. Here, we propose a method that addresses these limitations; improving on several aspects of the tractography segmentation process in order to reduce the overall variability in bundle segmentations. This method results in more repeatable, reproducible, and anatomically accurate identification of white matter pathways, which can significantly enhance the sensitivity and specificity of tractography studies.

2   Methods

In this section, we detail our proposed tractography bundle segmentation, named BundleSeg. BundleSeg is a combined procedure using an iterative registration technique (2.1) that leverages an existing bundle distance measure (2.2) and a newly proposed streamline search algorithm (2.3). This integrated approach allows an efficient alignment of tractograms and segmentation of streamlines that presents strong geometric similarity to well-established brain fascicles. By incorporating these elements, our method aims to enhance both accuracy and reproducibility in tractography bundle segmentation.

While our template registration and segmentation strategy share similarities with existing methods such as RecoBundles [8] and RecoBundlesX [14], they differ at a critical step aiming to discard streamlines that are too dissimilar to the atlas. In the standard RecoBundles framework, this pruning step is performed exclusively on clusters of streamlines, which only approximates a complete streamline-to-streamline comparison. On the contrary, the pruning operation in BundleSeg is performed using an exact distance search over all streamlines. This streamline search method is described in the Sect. 2.3.

2.1 Global and Local Registration Procedure

The first step is to obtain a coarse registration between the native diffusion space of a subject and an atlas of bundles (generated in MNI-152). This step is performed using ANTs linear registration [1] on anatomical images. An exact alignment is not necessary since the next step was designed to more closely align WM pathways. After this first coarse full brain registration, each white matter bundle of the atlas is further aligned with an iterative procedure alternating between searching for the closest and most similar streamlines in the tractogram and a streamline registration algorithm. This procedure allows to gradually improve the alignment and find more accurately trajectories that are similar to the desired bundle, comparable to the streamline-based linear registration (SLR) [9].

2.2 Streamlines Distance

The minimum average direct-flip (MDF) distance is a reliable way to compute the distance between streamlines. This distance is employed to estimate how similar a streamline is to the template atlas. It has been used in various algorithms related to tractography, such as clustering, classification and outlier detection [7,9,13,21]. When two streamlines (U,W) have the same orientation, the MDF is equivalent to averaging the Euclidean distance along all 'm' points $(\mathbf{u}_i, \mathbf{w}_i)$ of those two curves: $\text{dist}(U, W) = \sum_{i=1}^{m} ||\mathbf{u}_i - \mathbf{w}_i||_2$.

2.3 Streamlines Search

Typically, the MDF is used to find the distance between reference streamlines (e.g. a white matter atlas) and another set of streamlines (e.g. a whole brain tractogram). However, when millions of streamlines are involved in the context of an exhaustive comparison (to find the nearest neighbor), even the most efficient distance computation will result in astronomically high computation time. Computing all possible pairs of distances would result in the MDF being computed trillions of times $\sim O(n^2)$.

To avoid this, a K-D tree was adapted to search for nearby tractography streamlines. The resulting space partitioning tree drastically reduces the amount of computation required to find similar streamlines within a specific radius $\sim O(n \cdot log(n))$. This was possible by exploiting the Fast Streamline Search (FSS) mathematical framework recently developed to compute distances only within a maximum distance in the space of streamline [17].

As such, BundleSeg can accurately compute the bundle distance, from the WM template to all streamlines in the tractogram, at every step of the iterative procedure. This allows to avoid an approximate intermediate clustering step (in RecoBundles), or the need to better estimate the pruning distance using multiple execution (strategy from RecoBundlesX) to achieve an increased reliability. Resulting in an reproducible streamline segmentation method, with a single parameter (the maximum radius), that can exploit any existing tractography atlas.

The proposed segmentation procedure can be summarized by these four steps: 1) registering the subject anatomical image to the WM template reference image, 2) searching for all streamlines that resemble the bundles of interest using the resulting search tree based on the FSS framework, 3) refining the alignment of each bundle independently through SLR, and 4) iteratively repeating steps 2 and 3 while decreasing search radius until reaching the desired distance threshold.

The resulting search tree along with distance computation functions are available in Dipy. In addition, a complete segmentation pipeline improved with an exact pruning step is also available at this repository.

3   Experiments

We designed experiments in order to quantify the repeatability and reproducibility of the proposed white matter bundle segmentation algorithm, from full brain tractograms. The repeatability was assessed using a run-rerun of each algorithm on the same set of streamlines directly. The reproducibility was assessed using a scan-rescan dataset and computing the entire process end-to-end from different scans of the same subject.

3.1   Dataset and Template

For the evaluation, we employed 43 subjects at two timepoints (scan-rescan) from the Human Connectome Project (HCP) [20]. Full brain probabilistic tractograms were reconstructed using both classical local tractography and particle filtering tractography [10], implemented in Dipy [6]. Streamlines were generated using a WM seeding approach following fiber orientation distribution function (fODF) [4,18]. Preprocessing steps included DW-MRI denoising (MRtrix) [19], brain extraction and tissues classification (FSL-BET, FSL-FAST) [27]. It is important to note that the HCP dataset has T1-weighted images already aligned to the subject's DW-MRI space. To facilitate the analysis, all subjects' resulting streamlines were aligned to the MNI-152 space (ICBM 2009c nonlinear symmetrical) T1-weighted average [5] using ANTs linear registration (antsRegistrationSyNQuick.sh) [1].

The WM bundle atlas selected for this segmentation comparison encompasses 48 bundles aligned in MNI-152 space (ICBM 2009c nonlinear symmetric) from a population average based on HCP & UKBioBank. This template was obtained through Zenodo, developed along RecoBundlesX [14]. A subset of well-known bundles were used for this evaluation: Arcuate Fasciculus (AF), Corpus Callosum frontal (CC_Fr_2) and central portion (CC_Pr_Po), Cingulum (CG), Inferior Fronto-Occipital Fasciculus (IFOF), Inferior Longitudinal Fasciculus (ILF), Pyramidal Tract (PYT), Superior Longitudinal Fasciculus (SLF).

3.2   Evaluation

To assess the repeatability and reproducibility of the proposed tractography bundle segmentation method, a total of two full brain tractograms were computed for each 43 HCP scan-rescan subjects, one at each session. This was done to determine the scan-rescan reproducibility for each compared method. In addition, the run-rerun variability was estimated by executing each bundle segmentation algorithm twice using identical streamlines and parameters, but with a distinct random number generator.

Multiple measures were used to evaluate each bundle segmentation approach: the bundle volume, the number of streamlines, and the average streamline length. To estimate the variability, these measures were compared in both run-rerun and scan-rescan segmentations using an absolute difference (L1-norm) averaged over all subjects, along with the standard deviation ($\pm$).

Volumetric Dice coefficient was computed for both run-rerun and scan-rescan, to evaluate the overall volume similarity, as well as voxel overlap and overreach. For non-overlapping voxels, the average distance to the nearest corresponding voxel was computed, describing the "adjacency" distance between two segmentations. Compared to the overreach percentage in the Dice coefficient, this describes how distant on average are two successive bundle extractions.

BundleSeg computation time was compared to the standard RecoBundles as well as a multi-parameters RecoBundles with atlas fusion (RecoBundlesX) algorithms, using an Intel Skylake 6148 at 2.4 GHz.

4   Results

4.1   Computation Time

Computation times per subject for each method were: 18.52 ± 3.09 min (RB), 124.85 ± 21.60 min (RBX), 8.10 ± 1.52 min (BundleSeg, proposed). BundleSeg achieves a remarkable increase in reproducibility and overall quality compared to RecoBundles (RB) for an execution twice (2.3x) as fast. BundleSeg surpasses RecoBundlesX (RBX) in reproducibility and overall quality with an execution approximately 15 times faster. Our proposed method is faster than both baselines while offering more reproducible and better quality segmentations in both run-rerun and scan-rescan settings, qualitative and quantitative analysis of segmentation results are presented below. The computation time includes loading, clustering, streamlines search and distance computation, and finally saving. This loading, clustering and saving are the bulk ($\sim$70%) of the computation time for BundleSeg.

4.2   Qualitative Comparison, Segmentation Overlap and Overreach

Figure 1 visualizes streamlines and volumetric results for the run-rerun analysis for RecoBundles (RB), RecoBundlesX (RBX), and the proposed method (BundleSeg) from a single randomly chosen subject. Areas of overlap (in orange) and difference (in pink) are shown for each bundle and each algorithm for both streamlines and voxel-wise. While all algorithms (for all pathways) show similar locations, shape, and size of pathways when rerun on the same data, RB often recognizes and segments very different streamlines when run twice on the same algorithm, which can result in different estimated pathway volumes. This run-rerun variation is considerably reduced by RBX, and nearly non-existent with BundleSeg.

Figure 2 shows the volumetric overlap-overreach in a scan-rescan setting, with two independent segmentations from different DW-MRI acquisitions. Again, all algorithms result in visually similar scan-rescan segmentations, but there are variations, particularly at the edge of bundles, with more variation in RB, followed by RBX, and BundleSeg.

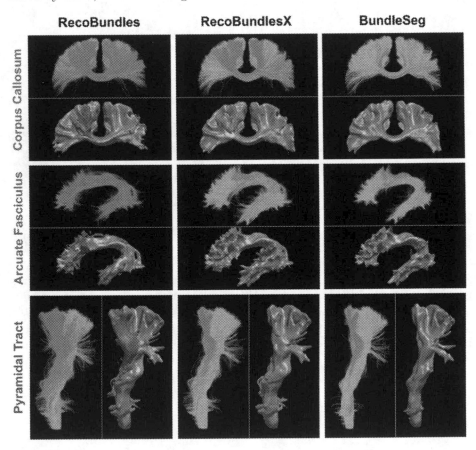

Fig. 1. Run-rerun visual comparison of two successive segmentations with the exact same streamlines (different random seeds), between two baselines and the proposed method over 3 bundles. Bundle streamlines and voxels overlap are displayed in orange (consistent in both run-rerun segmentations) and overreach in pink (inconsistent). (Color figure online)

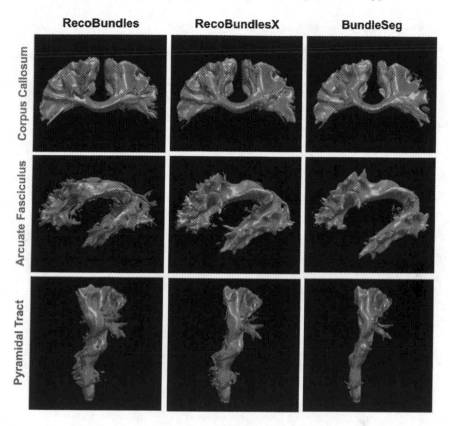

Fig. 2. Scan-rescan visual comparison of two independently reconstructed tractograms (from distinct acquisitions), between two baselines and the proposed method on three bundles. Bundle volume overlap is displayed in orange (consistent in both scan-rescan segmentations) and overreach in pink (inconsistent). (Color figure online)

4.3 Quantitative Results, Reproducibility and Variability

Figure 3 shows quantitative run-rerun and scan-rescan results, for all three methods, where several trends are apparent. First, for almost every pathway the Dice coefficient shows that BundleSeg outperforms both the standard RecoBundles framework, as well as RecoBundlesX for most bundles (while equal for others), in both run-rerun and scan-rescan. In run-rerun, BundleSeg results in a near-perfect volumetric match (> 0.95) in all tested pathways. Scan-rescan shows an expected decreased volume overlap for all methods when compared to run-rerun results. Next, BundleSeg has the lowest adjacency distances in both run-rerun and scan-rescan, and RBX is second. Importantly, adjacency distance is around one or two voxels (mm) on average, for all algorithms. Finally, differences in volume can largely vary when using RB, even on the same tractogram. Repeatability and reproducibility of volume can be strongly improved through RBX, and further enhanced with BundleSeg.

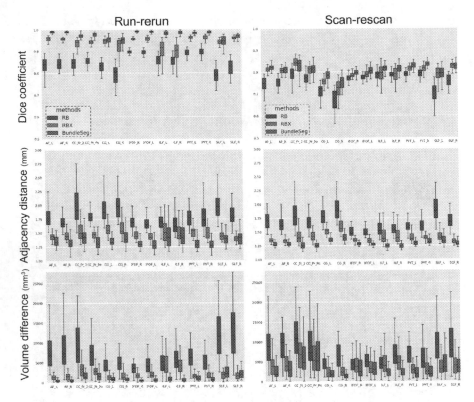

Fig. 3. Comparison of agreement in both run-rerun (left) and scan-rescan (right) for 3 measures: Dice coefficient (top row, score of 1 is a perfect match) Adjacency, millimetric distance for non-overlapping voxel (middle row, lower is better with a minimum of 1 mm) and Difference in volume in mm^3 (bottom row, 0 is better). We compared the standard RecoBundles framework, the multi-parameters with atlas fusion RecoBundles framework (RecoBundlesX) and our proposed method. A sample of representative major WM pathways were selected to showcase the agreement measures. The proposed BundleSeg surpasses RecoBundles and RecoBundlesX framework is nearly all bundles, for all 3 agreement measures for both the run-rerun and scan-rescan. The average scores are improved, and the standard deviation is decreased.

5 Discussion

In this work, we introduced a bundle segmentation algorithm that overcomes challenges associated with existing manual and automated machine learning-based segmentation techniques. This method, named BundleSeg, resulted in a significant computational speedup, and improved repeatability and reproducibility compared to current state-of-the-art methods.

While RecoBundlesX has much better reproducibility than its predecessor RecoBundles, it requires multiple comparisons, resulting in higher computation time. On the other hand, the proposed method is 2x and 15x faster than RecoBundles and RecoBundlesX respectively. This can be attributed to the fast streamlines search making the distance computation and pruning an order of magnitude faster than the clustering approach of RecoBundles and RecoBundlesX.

The proposed method displayed an improvement in both scan-rescan reproducibility compared to existing approaches. This can be observed qualitatively in Figs. 1 and 2, and quantitatively in all graphs from Fig. 3. BundleSeg has a significantly higher Dice coefficient, indicating a greater agreement between the repeated segmentations, more specifically in run-rerun where it has a near-perfect Dice along with a very good adjacency. For run-rerun comparison, a "Streamline Dice" coefficient was also computed describing the overlap of streamlines (as the element of comparison instead of voxels); but was not included since results were similar to "Voxel Dice" results.

Furthermore, our results exhibited a notable reduction in volume variability. This lower same-subject absolute volume difference in WM pathway segmentations indicates that our method consistently extracted a similarly shaped bundle. The reduced variability indicates the stability and reliability of our approach, minimizing the influence of random initialization or other factors that could introduce variability.

All our results suggest that the proposed algorithm reduces the inherent variability associated with tractography bundle segmentation, enabling more consistent and reproducible results across multiple sessions. While the reduced Dice alignment in scan-rescan is lower, part of it might be caused by the tractography reconstruction variability.

Future work should include more comparison to other segmentation methods, based on traditional classification and deep learning. Second, it remains to be seen whether the higher precision and improved reproducibility lead to more sensitive detection of white matter changes in disease and disorder. And finally, it is worth exploring how this exhaustive and fast streamline search can improve longitudinal analysis on subjects with multiple timepoints.

6   Conclusion

In this work, we proposed a reliable and robust approach for extracting white matter pathways. The novelty of BundleSeg resides in a precise search algorithm, which efficiently identifies all relevant tractography streamlines corresponding to specific white matter bundles. Using an exact and exhaustive streamline radius search, instead of an approximation, ensures that the segmentation process is comparable to previous work while significantly improving speed and reliability. Furthermore, combining it with an iterative registration results in an overall higher reproducibility and lower variability. This improved stability provided by our method has important implications for neuroimaging studies, enabling researchers to obtain more reliable and robust results when investigating white matter connectivity patterns and their associations with various clinical or cognitive variables.

Conflict of Interest. We have no conflict of interest to declare.

References

1. Avants, B.B., Epstein, C.L., Grossman, M., Gee, J.C.: Symmetric diffeomorphic image registration with cross-correlation: evaluating automated labeling of elderly and neurodegenerative brain. Med. Image Anal. **12**(1), 26–41 (2008)
2. Bertò, G., et al.: Classifyber, a robust streamline-based linear classifier for white matter bundle segmentation. Neuroimage **224**, 117402 (2021)
3. Catani, M., et al.: Symmetries in human brain language pathways correlate with verbal recall. Proc. Natl. Acad. Sci. **104**(43), 17163–17168 (2007)
4. Descoteaux, M., Deriche, R., Knosche, T.R., Anwander, A.: Deterministic and probabilistic tractography based on complex Fibre orientation distributions. IEEE Trans. Med. Imaging **28**(2), 269–286 (2008)
5. Fonov, V., et al.: Unbiased average age-appropriate atlases for pediatric studies. Neuroimage **54**(1), 313–327 (2011)
6. Garyfallidis, E., et al.: Dipy, a library for the analysis of diffusion MRI data. Front. Neuroinformatics **8**, 8 (2014)
7. Garyfallidis, E., Brett, M., Correia, M.M., Williams, G.B., Nimmo-Smith, I.: Quickbundles, a method for tractography simplification. Front. Neurosci. **6**, 175 (2012)
8. Garyfallidis, E., et al.: Recognition of white matter bundles using local and global streamline-based registration and clustering. Neuroimage **170**, 283–295 (2018)
9. Garyfallidis, E., Ocegueda, O., Wassermann, D., Descoteaux, M.: Robust and efficient linear registration of white-matter fascicles in the space of streamlines. Neuroimage **117**, 124–140 (2015)
10. Girard, G., Whittingstall, K., Deriche, R., Descoteaux, M.: Towards quantitative connectivity analysis: reducing tractography biases. Neuroimage **98**, 266–278 (2014)
11. O'Donnell, L.J., Westin, C.F.: Automatic tractography segmentation using a high-dimensional white matter atlas. IEEE Trans. Med. Imaging **26**(11), 1562–1575 (2007)
12. Oishi, K., et al.: Human brain white matter atlas: identification and assignment of common anatomical structures in superficial white matter. Neuroimage **43**(3), 447–457 (2008)
13. Olivetti, E., Berto, G., Gori, P., Sharmin, N., Avesani, P.: Comparison of distances for supervised segmentation of white matter tractography. In: 2017 International Workshop on Pattern Recognition in Neuroimaging (PRNI), pp. 1–4. IEEE (2017)
14. Rheault, F.: Analyse et Reconstruction de Faisceaux de la Matière Blanche. Université de Sherbrooke, Computer Science (2020)
15. Rheault, F., et al.: The influence of regions of interest on tractography virtual dissection protocols: general principles to learn and to follow. Brain Struct. Funct. **227**(6), 2191–2207 (2022)
16. Schilling, K.G., et al.: Tractography dissection variability: what happens when 42 groups dissect 14 white matter bundles on the same dataset? Neuroimage **243**, 118502 (2021)
17. St-Onge, E., Garyfallidis, E., Collins, D.L.: Fast streamline search: an exact technique for diffusion MRI tractography. Neuroinformatics **20**(4), 1093–1104 (2022)

18. Tournier, J.D., Calamante, F., Connelly, A.: Robust determination of the Fibre orientation distribution in diffusion MRI: non-negativity constrained super-resolved spherical deconvolution. Neuroimage **35**(4), 1459–1472 (2007)
19. Tournier, J.D., et al.: Mrtrix3: a fast, flexible and open software framework for medical image processing and visualisation. Neuroimage **202**, 116137 (2019)
20. Van Essen, D.C., et al.: The WU-Minn human connectome project: an overview. Neuroimage **80**, 62–79 (2013)
21. Visser, E., Nijhuis, E.H., Buitelaar, J.K., Zwiers, M.P.: Partition-based mass clustering of tractography streamlines. Neuroimage **54**(1), 303–312 (2011)
22. Wakana, S., et al.: Reproducibility of quantitative tractography methods applied to cerebral white matter. Neuroimage **36**(3), 630–644 (2007)
23. Wassermann, D., et al.: On describing human white matter anatomy: the white matter query language. In: Mori, K., Sakuma, I., Sato, Y., Barillot, C., Navab, N. (eds.) MICCAI 2013. LNCS, vol. 8149, pp. 647–654. Springer, Heidelberg (2013). https://doi.org/10.1007/978-3-642-40811-3_81
24. Wasserthal, J., Neher, P., Maier-Hein, K.H.: TractSeg-fast and accurate white matter tract segmentation. Neuroimage **183**, 239–253 (2018)
25. Zhang, F., Karayumak, S.C., Hoffmann, N., Rathi, Y., Golby, A.J., O'Donnell, L.J.: Deep white matter analysis (DeepWMA): fast and consistent tractography segmentation. Med. Image Anal. **65**, 101761 (2020)
26. Zhang, Y., et al.: Atlas-guided tract reconstruction for automated and comprehensive examination of the white matter anatomy. Neuroimage **52**(4), 1289–1301 (2010)
27. Zhang, Y., Brady, M., Smith, S.: Segmentation of brain MR images through a hidden Markov random field model and the expectation-maximization algorithm. IEEE Trans. Med. Imaging **20**(1), 45–57 (2001)

Automated Mapping of Residual Distortion Severity in Diffusion MRI

Shuo Huang[1,2], Lujia Zhong[1,3], and Yonggang Shi[1,2,3]($\boxtimes$)

[1] Stevens Neuroimaging and Informatics Institute, Keck School of Medicine, University of Southern California (USC), Los Angeles, CA 90033, USA
yonggans@usc.edu
[2] Alfred E. Mann Department of Biomedical Engineering, Viterbi School of Engineering, University of Southern California (USC), Los Angeles, CA 90089, USA
[3] Ming Hsieh Department of Electrical and Computer Engineering, Viterbi School of Engineering, University of Southern California (USC), Los Angeles, CA 90089, USA

Abstract. Susceptibility-induced distortion is a common artifact in diffusion MRI (dMRI), which deforms the dMRI locally and poses significant challenges in connectivity analysis. While various methods were proposed to correct the distortion, residual distortions often persist at varying degrees across brain regions and subjects. Generating a voxel-level residual distortion severity map can thus be a valuable tool to better inform downstream connectivity analysis. To fill this current gap in dMRI analysis, we propose a supervised deep-learning network to predict a severity map of residual distortion. The training process is supervised using the structural similarity index measure (SSIM) of the fiber orientation distribution (FOD) in two opposite phase encoding (PE) directions. Only b0 images and related outputs from the distortion correction methods are needed as inputs in the testing process. The proposed method is applicable in large-scale datasets such as the UK Biobank, Adolescent Brain Cognitive Development (ABCD), and other emerging studies that only have complete dMRI data in one PE direction but acquires b0 images in both PEs. In our experiments, we trained the proposed model using the Lifespan Human Connectome Project Aging (HCP-Aging) dataset ($n = 662$) and apply the trained model to data ($n = 1330$) from UK Biobank. Our results show low training, validation, and test errors, and the severity map correlates excellently with an FOD integrity measure in both HCP-Aging and UK Biobank data. The proposed method is also highly efficient and can generate the severity map in around 1 s for each subject.

Keywords: Susceptibility-induced distortion · Residual distortion severity map · B0 images

Y. Shi—This work is supported by the National Institute of Health (NIH) under grants R01EB022744, RF1AG077578, RF1AG056573, RF1AG064584, R21AG064776, U19AG078109, and P41EB015922.

M. Karaman et al. (Eds.): CDMRI 2023, LNCS 14328, pp. 58–69, 2023.
https://doi.org/10.1007/978-3-031-47292-3_6

1   Introduction

Susceptibility-induced distortion is a common artifact in diffusion MRI (dMRI), which causes stretching or pressing of brain structures along the phase encoding (PE) direction, particularly in the brainstem, orbitofrontal and temporal cortical areas [1–3]. While various methods have been proposed for distortion correction, residual distortions often persist and pose a significant yet unaccounted for challenge in common tasks such as fiber tracking (Fig. 1). To better inform connectivity analysis based on dMRI, there is thus a current gap in automatically predicting the severity of residual distortion across different brain regions, which would greatly enhance the rigor in connectome research by enabling the consideration of this uncertainty in downstream statistical modeling.

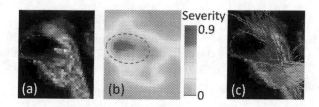

Fig. 1. An illustration of the impact of residual distortion on connectivity analysis in the brainstem area of one HCP-Aging subject. (a) shows the first spherical harmonics component of the FOD image, which corresponds to the $L = 0$ component in Eq. 1. (b) displays the residual distortion severity map generated from FOD-based features. (c) shows the severe residual distortion prevents the successful fiber tracking through the pons (dashed red ellipse).

Previous methods for distortion correction are mainly based on image registration techniques [4]. Based on the assumption that distortions have opposite directions and the same magnitude in the two opposing PEs, image registration using b0 images of opposite PEs have been proposed and incorporated into popular tools such as *Topup* in FSL [5,6]. For data from Human Connectome Projects (HCP) [7–11] and related studies, complete dMRI were acquired in two opposite PEs. Recently, more informative features such as the fiber orientation distribution (FOD) [12] from both PEs were proposed to further remove the distortion artifacts in dMRI, but residual distortions still persist in varying degrees across subjects [13–15]. Nevertheless, with FOD-based features from opposite PEs, it is very straightforward to locate areas with significant residual distortions (Fig. 2) and hence generate a map of distortion severity. For many large-scale studies including UK-Biobank [16–18] and ABCD [19], and many other clinical studies, however, complete dMRI data are typically limited to one PE and only b0 images are collected from both PEs due to time constraints. This practically limits our ability to use FOD-based features for distortion correction and severity analysis. For these studies, it is desirable that we can generate the severity map of residual distortions based only on b0 images.

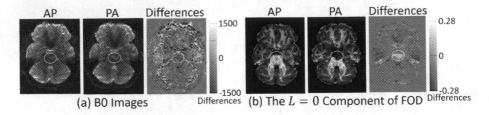

Fig. 2. The presence of residual distortion in the pons area can be better observed from (b) the difference of the first component ($L = 0$ component in Eq. 1) of FOD images between two PEs than (a) the differences of b0 images. The red ellipses mark the same position on brainstem that has high FOD differences.

However, current tools for dMRI quality control cannot yet address the residual distortion map generation. For example, Gaussian process (GP) based prediction is a widely accepted method used in the Eddy tool [20,21] of FSL, which measures the error after the correction of eddy currents and head motion. Although the GP based prediction method performs well on the artifacts that have high variability between each volume of the dMRI data, such as eddy currents and head motion, it doesn't work well for the residual susceptibility-induced distortion that affects all volumes of dMRI data to a similar degree. Figure 3(b1)– (c4) show maps of the GP based prediction error for the data in Fig. 2. For AP data, the residual distortion at brainstem region cannot be extracted using the GP based prediction error maps. For PA data, the high-error regions at the brainstem have different shape, location and error value across different volumes of dMRI. Due to the inconsistent results in different volumes, the GP based prediction is not suitable for measuring the residual distortions that are constant across all volumes.

For the automated generation of residual distortion severity map from only b0 images of two PEs, we will develop in this work a novel supervised deep learning method. Using HCP style dMRI data that allows the calculation of FOD-based features in two PEs, we train a network that learns to predict a ground truth FOD-based severity map from only b0 related feature images, which can then be applied to more widely available data from studies such as UK-Biobank and ABCD. In our experiments, we use the HCP-Aging [9,10] data to train the network and successfully test the method on both HCP-Aging and UK-Biobank data. We show that our method is highly efficient and generates reliable prediction of distortion severity that correlates excellently with a measure of FOD-based white matter integrity.

2   Method

In this study, a supervised deep-learning model was utilized to generate the residual distortion severity map for dMRI images. Initially, the model was trained on datasets containing dMRI images in two opposing PE directions. After that,

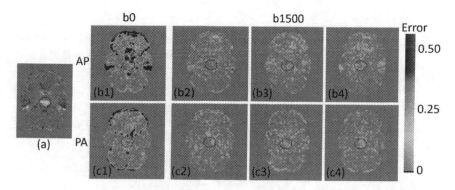

Fig. 3. Maps of GP based prediction error for the data in Fig. 2. (a) shows the differences between the $L = 0$ components of FODs in AP and PA directions. (b1)–(b4) and (c1)–(c4) show the loss maps for AP and PA dMRI data, respectively. The dMRI volumes of (b2) (c2), (b3) (c3) and (b4) (c4) have gradient directions close to $(1, 0, 0)$, $(0, 1, 0)$ and $(0, 0, 1)$, respectively. The high residual distortion region on brainstem is marked by the red ellipse in (a), and the same position is marked out in (b1)–(c4).

the model was applied to and tested on datasets containing dMRI images in only one PE direction, and only b0 images in the opposite PE direction.

2.1 Ground Truth Calculation

The ground truth for the training of our model is the residual distortion severity map calculated by the FODs of dMRI data in two opposite PE directions. The residual distortion severity map is calculated by the structural similarity index measure (SSIM) map [22,23] of the first 6 components of the FODs.

The FOD is calculated using dMRI images after pre-processing, including susceptibility-induced distortion correction and correction of artifacts of eddy currents and head motion. It reflects fiber orientation and provides detailed information derived from all gradient directions of dMRI images. The FOD is calculated using the method proposed in Ref. [24], where the authors defined a constrained minimization framework using an energy function that consists of a data fidelity term and a regularization term for the sparsity of the FOD, and then they developed a coordinate decent algorithm to solve the FOD reconstruction problem. The FOD is represented by spherical harmonics (SPHARM) up to the order L [24], which is defined by Eq. 1:

$$f(p) = \sum_{l,m} s_l^m \varPhi_l^m(p), \forall p \in \mathbb{S} \tag{1}$$

where p is a fiber direction on the unit sphere $\mathbb{S}$, s_l^m represents the SPHARM coefficient at order $l = 0, 2, 4, \ldots, L$, and $\varPhi_l^m$ denotes the $m - th$ real SPHARM basis. We use $L = 2$ in the generation of ground truth, since it is proven that $L = 2$ can provide enough information for distortion detection [3]. The FOD contains $J = 6$ components for $L = 2$, where $J = (L + 1)(L + 2)/2$.

The SSIM map in the $j - th$ component of FOD (denoted as $FOD_{pos}(j)$ and $FOD_{neg}(j)$ for the FOD in the positive and negative PE directions, respectively) is calculated by Eq. 2 using voxels in each voxel's $3 \times 3 \times 3$ neighborhood. It is a voxel-level measurement of the similarity of the $j - th$ FOD components in the positive and negative PE directions:

$$SSIM(j) = \frac{2\mu_{FOD_{pos}(j)}\mu_{FOD_{neg}(j)} + C_1}{\mu^2_{FOD_{pos}(j)} + \mu^2_{FOD_{neg}(j)} + C_1} \cdot \frac{2\sigma_{FOD_{pos}(j)}\sigma_{FOD_{neg}(j)} + C_2}{\sigma^2_{FOD_{pos}(j)} + \sigma^2_{FOD_{neg}(j)} + C_2}$$
$$\cdot \frac{cov(FOD_{pos}(j), FOD_{neg}(j)) + C_3}{\sigma_{FOD_{pos}(j)}\sigma_{FOD_{neg}(j)} + C_3}$$

$$(2)$$

where μ and θ correspond to the voxel sample mean and variance of the value of the $j - th$ component of FOD coefficients of each voxel in $FOD_{pos}(j)$ and $FOD_{neg}(j)$ within its $3 \times 3 \times 3$ neighborhood window, respectively. The term $cov(FOD_{pos}, FOD_{neg})$ represents the cross-covariance between voxels in each voxel's $3 \times 3 \times 3$ neighborhood window in the $j - th$ component of FOD_{pos} and FOD_{neg}. C_1, C_2 and C_3 are 3 small constants to avoid zero-division.

Compared with mean squared error (MSE) and local cross-correlation coefficients (LCC) parameters, SSIM can more comprehensively interpret the similarity of images and is more robust to intensity variations of FOD in different regions. As shown in Eq. 2, SSIM consists of three parts. The first part computes the difference in images' pixel values; the second part computes the difference in image contrast, so that SSIM is sensitive to areas with low FOD values; the third part computes the similarity of image structure via LCC.

The residual distortion severity of each voxel in the severity map, $d(x, y, z)$, is calculated using the normalized $L2 - norm$ value of the SSIM of each FOD component:

$$d(x, y, z) = \sqrt{\sum_{j=1}^{6} (1 - SSIM(j)(x, y, z))^2 / 6} \qquad (3)$$

Since the range of SSIM is $[-1, 1]$, where 1 means that two FODs have the same value, and -1 means the they have opposite signs and the same amplitude, the term $(1 - SSIM(j))$ is used to ensure the severity value is non-negative. Higher d means more severe residual distortion at that voxel.

In Fig. 4, an example of a residual distortion severity map d is shown. The yellow and red regions are areas where residual distortion is located. Higher value on the map indicates more severe residual distortion. The residual distortion at temporal lobe can also be detected, even that the FOD value at this region is much lower than the brainstem.

2.2   Network Architecture

Figure 5 shows the network architecture of our proposed method. The input of our method are distortion distribution map obtained from the distortion correction method, as well as only b0 images in two PE directions before and after

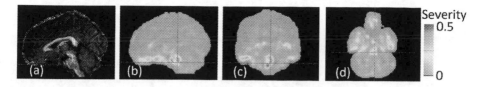

Fig. 4. Residual distortion severity map. (a) is the FOD image ($L = 0$). (b) - (d) show the residual distortion maps in a sagittal, coronal and axial slice, respectively.

being implemented the distortion correction and eddy currents and head motion correction methods. Therefore, the models trained in this way can be applied to datasets comprising dMRI data in only one PE direction.

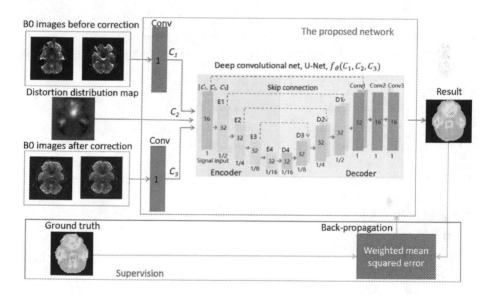

Fig. 5. Network architecture of the proposed method.

To provide the network with more information to facilitate learning, initially, we employ two convolution layers with a kernel size of $3 \times 3 \times 3$ to extract the correspondence between the b0 images in the two PE directions. Additionally, the distortion distribution map is included to provide the network with more insights into the distribution of distortion. It is noteworthy that residual distortion primarily occurs in areas with high distortion, and the distortion distribution map assists the network in identifying these regions.

B0 images appear dissimilar across different datasets. Therefore, we utilize a b0 image normalization to scale the b0 images such that the mean gray-level value of the voxels in the brain is set to 0.5 using Eq. 4:

$$b0_{norm} = r \times b0, \; where \; r = 0.5 / \underset{\substack{(x,y,z) \; in \; brain}}{mean} b0(x,y,z) \tag{4}$$

Moreover, image up- and down-sampling methods are employed to make all data used in training and test have the same size and voxel spacing.

To extract information from the inputs, we utilize the U-Net architecture. U-Net [3,25,26] is known for its ability to efficiently extract information from various resolutions and frequency ranges. In our proposed method, U-Net uses convolutional layers with a kernel size of $3 \times 3 \times 3$ and a Leaky Rectified Linear Unit (ReLU) activation layer with a parameter of 0.2. In order to prevent overfitting, the $L2$ penalty is implemented in all convolution layers.

In the training process, we utilize a weighted mean squared error e to measure the differences between each voxel in the ground truth d and the predicted severity map d_p, shown in Eq. 5:

$$e(x, y, z) = ((d(x, y, z) - d_p(x, y, z)) \times (1 + d(x, y, z)))^2 \tag{5}$$

Notably, by the term $(1 + d(x, y, z))$, we use the value of the ground truth, $d(x, y, z)$, as the weight to enlarge the differences in regions with severe residual distortion. This is particularly useful in this study since such areas are relatively small compared to the overall volume size.

2.3   Datasets, Model Training and Validation

The proposed model is trained on the HCP-Aging dataset, and being tested on UK Biobank dataset. These datasets both collect dMRI images in anterior-posterior (AP) and posterior-anterior(PA) PE directions, and the subjects in these two datasets have a similar age range. Each scan in HCP-Aging has 99 gradient directions over 2 shells with b values of 1500 and 3000 s/mm^2, while each scan in UK Biobank dataset has 100 gradient directions over two shells with b values of 1000 and 2000 s/mm^2. Table 1 shows information of data used in this study. The size and spacing of the HCP-Aging data is interpolated and cropped to those of UK Biobank. The *Topup* [5] is employed to correct susceptibility-induced distortion, and *Eddy* in *FSL* is utilized to correct artifacts caused by eddy currents and head motion.

Table 1. Information of HCP-Aging and UK Biobank datasets

Datasets	Set name	Number of subjects	Image size (voxel)	Resolution (mm^3)
HCP-Aging	Training	500	$140 \times 140 \times 92$	$1.5 \times 1.5 \times 1.5$
HCP-Aging	Test	162	$140 \times 140 \times 92$	$1.5 \times 1.5 \times 1.5$
UK Biobank	Application	1330	$104 \times 104 \times 72$	$2 \times 2 \times 2$

Our proposed method was trained for a total of 1500 epochs with 25 steps per epoch by Anaconda3 using Python3, which requires GPU memory for 2 G bytes. Stochastic gradient descent was employed, where one training subject was randomly selected per step. Our optimization used the ADAM optimizer with a learning rate of 10^{-4}, and an $L2$ penalty weight of 10^{-5}.

We utilized 5-fold cross-validation [27,28] during the training process to choose the value of hyperparameters including learning rate, $L2$ penalty rate and the number of epochs. The 662 HCP-Aging subjects were divided into two sets: 500 subjects for training and 162 subjects for independent testing. For the 500 subjects in the first set, we performed 5-fold cross-validation for training and hyperparameter tuning. More specifically, 500 subjects were randomly formed into 5 subsets, with 100 subjects in each subset. Then we trained 5 models for each set of hyperparameters. For each model, we chose 4 subsets to train the model and the rest was served as the validation set, and then each subset was used for validation in one model. The hyperparameters were selected to minimize the average loss in all five validation subsets to avoid overfitting. After the cross-validation, a model was defined with the best set of hyperparameters, trained with all 500 data and tested on the test set.

Compared with a random split of training and validation sets, 5-fold cross-validation can end in a more robust and reliable model, since it enables the model to make full use of all 500 data in the training set and generalizes the performances of the model on multiple validation subsets.

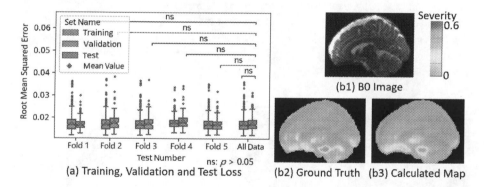

Fig. 6. The training, validation, and test loss of the HCP-Aging dataset (a) and a case in the test set that presents the ground truth and the calculated map (b). In figure (a), "ns" indicates that the difference is not statistically significant with $p > 0.05$.

3 Results

3.1 Results on HCP-Aging Dataset

The root mean squared error (RMSE) in the training, validation, and test sets on HCP-Aging dataset of the model with the selected hyperparameters is shown in Fig. 6. In Fig. 6(a), the maximum RMSE for the training, validation and test sets is less than 0.04, and the mean value of the RMSE is less than 0.02. Student's t-test result shows no significant differences between the training set and test set. Similarly, the RMSE shows that all validation sets have no significant difference with the test set. This indicates that the model is robust and not over-fitted to the training set. In Fig. 6(b), we visualized one data from the test set in Fig. 6(a). Compared with the ground truth, the calculated residual distortion severity map has similar high-distortion regions and severity values.

3.2   Results on UK Biobank Dataset

We used the model trained on the HCP-Aging dataset to predict the severity of residual distortion in the UK Biobank dataset. To quantitatively evaluate our method's performance, we calculated the mean residual distortion severity and mean FOD value within a region of interest in the brainstem, which is well-known for severe residual distortions, shown in Fig. 7(a). Because the FODs are computed with compartment-based models, the mean FOD equals to the $L = 0$ component in Eq. 1, which reflects the intra-axonal volume fraction of the compartment model and hence the integrity of the FOD model [24]. A reduction of mean FOD could be due to white matter degeneration or distortion artifacts. Given that only healthy subjects were included in our experiments, we use the mean FOD as an independent surrogate measure of distortion artifacts. Figure 7 shows scatter plots of the mean FOD versus the mean severity of the two datasets. To locate the region of interest of each subject, we registered the *Topup* and *Eddy* results of b0 images in the AP direction to the FMRIB58 FA common space [29] using *Elastix* [30], and resampled the mean FOD to that space.

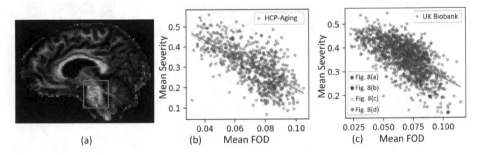

(a)   (b)   (c)

Fig. 7. The scatter map of the mean FOD and mean residual distortion severity in the brainstem. Figure (a) illustrates the brainstem region, while figures (b) and (c) show the scatter plots for data in the HCP-Aging and UK Biobank datasets, respectively.

The negative correlation between the mean severity and mean FOD of HCP-Aging dataset is shown in Fig. 7(b), and the Pearson correlation coefficient is –0.665. The similar correlation in the UK Biobank dataset is shown in Fig. 7(c), and the Pearson correlation coefficient is –0.646. Similar distribution and correlation in the two datasets prove that our method is reliable to predict residual distortion's severity. Figure 8 displays four examples, whose residual distortion severity gradually increase and their mean FOD values gradually decrease. The

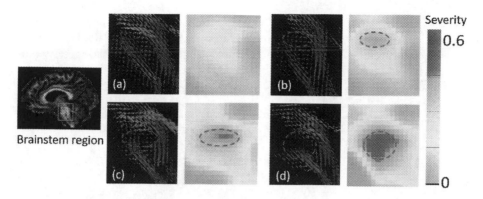

Fig. 8. The FOD and residual distortion severity maps of the brainstem for four cases in UK Biobank dataset. The FODs in the high distortion regions marked by the ellipse get more and more disorganized as residual distortion severity increases. The left subfigures of (a)–(d) show FODs, and the right subfigures are the residual distortion severity maps of each case.

FODs are more organized under low residual distortion severity, as shown in Fig. 8(a). For Figs. 8(b) to (d), as the residual distortion severity increases, the FODs are more and more disorganized.

4   Discussion and Conclusion

This study draws attention to an understudied but important question and proposes a supervised deep learning solution, which aimed at predicting the voxel-level residual distortion severity map of dMRI distortion correction results. The calculation time of our method is around one second and it avoids FOD calculation in testing, which makes it applicable to large-scale data analysis. Result shows that our method can reliably and efficiently generate the localized residual distortion map for both HCP-Aging and UK Biobank datasets.

The proposed method can efficiently and reliably map the voxel-level localized severity of the residual distortion for the preprocessed dMRI data. With the severity map, researchers can locate the voxels or brain regions that have high residual distortions which may change brain connectivity. They can strategically process or exclude those distorted regions, which are similar with pathological regions but caused by residual distortions, so as to mitigate negative impact on subsequent downstream connectivity and statistical analysis.

One limitation of our method is the limited number of datasets that has dMRI data in 2 PE directions, and these datasets are mandatory for training. Another limitation is that our method is not applicable to datasets with b0 images in only one phase encoding direction, such as the Alzheimer's Disease Neuroimaging Initiative (ADNI) [31].

In the future, we will further test the proposed method on additional datasets, such as the Lifespan Human Connectome Project Development (HCP-D) [32] and the ABCD Study. We will also focus on enhancing the accuracy of predicted residual distortion map and extending the application of predicting the severity of artifacts like off-resonance effects and subject movement [33, 34].

Acknowledgements. Authors thank Dr. Yuchuan Qiao from Fudan University for the kindly help and useful discussions. We also appreciate Mr. Sarthak Kumar Maharana from University of Southern California for the help in tractography. Shuo Huang wants to thank Miss Yi Liu from University of Southern California for the help in editing the grammar of this work.

References

1. O'Donnell, L.J., Westin, C.F.: An introduction to diffusion tensor image analysis. Neurosurg. Clin. **22**(2), 185–196 (2011)
2. Li, J., Shi, Y., Toga, A.W.: Mapping brain anatomical connectivity using diffusion magnetic resonance imaging: structural connectivity of the human brain. IEEE Signal Process. Mag. **33**(3), 36–51 (2016)
3. Qiao, Y., Shi, Y.: Unsupervised deep learning for fod-based susceptibility distortion correction in diffusion MRI. IEEE Trans. Med. Imaging **41**(5), 1165–1175 (2021)
4. Sotiras, A., Davatzikos, C., Paragios, N.: Deformable medical image registration: a survey. IEEE Trans. Med. Imaging **32**(7), 1153–1190 (2013)
5. Andersson, J.L., Skare, S., Ashburner, J.: How to correct susceptibility distortions in spin-echo echo-planar images: application to diffusion tensor imaging. Neuroimage **20**(2), 870–888 (2003)
6. Jenkinson, M., Beckmann, C.F., Behrens, T.E., Woolrich, M.W., Smith, S.M.: FSL. Neuroimage **62**(2), 782–790 (2012)
7. Van Essen, D.C., et al.: The human connectome project: a data acquisition perspective. Neuroimage **62**(4), 2222–2231 (2012)
8. Van Essen, D.C., et al.: The WU-MINN human connectome project: an overview. Neuroimage **80**, 62–79 (2013)
9. Bookheimer, S.Y., et al.: The lifespan human connectome project in aging: an overview. Neuroimage **185**, 335–348 (2019)
10. HCP-Aging Homepage. https://www.humanconnectome.org/study/hcp-lifespan-aging. Accessed 8 Mar 2023
11. HCLV Homepage. https://www.humanconnectome.org/study/crhd-human-connectomes-low-vision-blindness-and-sight-restoration. Accessed 4 Aug 2023
12. Qiao, Y., Sun, W., Shi, Y.: FOD-based registration for susceptibility distortion correction in brainstem connectome imaging. Neuroimage **202**, 116164 (2019)
13. Hsu, Y.C., Tseng, W.Y.I.: DACO: Distortion/artefact correction for diffusion MRI data. Neuroimage **262**, 119571 (2022)
14. Schilling, K.G., et al.: Distortion correction of diffusion weighted MRI without reverse phase-encoding scans or field-maps. PLoS ONE **15**(7), e0236418 (2020)
15. Duong, S.T., Phung, S.L., Bouzerdoum, A., Schira, M.M.: An unsupervised deep learning technique for susceptibility artifact correction in reversed phase-encoding epi images. Magn. Reson. Imaging **71**, 1–10 (2020)
16. Bycroft, C., et al.: The UK biobank resource with deep phenotyping and genomic data. Nature **562**(7726), 203–209 (2018)
17. UK Biobank Homepage. https://www.ukbiobank.ac.uk. Accessed 8 Mar 2023

18. Sudlow, C., et al.: UK biobank: an open access resource for identifying the causes of a wide range of complex diseases of middle and old age. PLoS Med. **12**(3), e1001779 (2015)
19. Bjork, J.M., Straub, L.K., Provost, R.G., Neale, M.C.: The ABCD study of neurodevelopment: identifying neurocircuit targets for prevention and treatment of adolescent substance abuse. Curr. Treatment Opt. Psychiat. **4**, 196–209 (2017)
20. Andersson, J.L., Sotiropoulos, S.N.: Non-parametric representation and prediction of single-and multi-shell diffusion-weighted MRI data using gaussian processes. Neuroimage **122**, 166–176 (2015)
21. Andersson, J.L., Sotiropoulos, S.N.: An integrated approach to correction for off-resonance effects and subject movement in diffusion MR imaging. Neuroimage **125**, 1063–1078 (2016)
22. Wang, Z., Simoncelli, E.P., Bovik, A.C.: Multiscale structural similarity for image quality assessment. In: The Thrity-Seventh Asilomar Conference on Signals, Systems & Computers, 2003, vol. 2, pp. 1398–1402. IEEE (2003)
23. Wang, Z., Bovik, A.C., Sheikh, H.R., Simoncelli, E.P.: Image quality assessment: from error visibility to structural similarity. IEEE Trans. Image Process. **13**(4), 600–612 (2004)
24. Tran, G., Shi, Y.: Fiber orientation and compartment parameter estimation from multi-shell diffusion imaging. IEEE Trans. Med. Imaging **34**(11), 2320–2332 (2015)
25. Ronneberger, O., Fischer, P., Brox, T.: U-net: convolutional networks for biomedical image segmentation. In: Navab, N., Hornegger, J., Wells, W.M., Frangi, A.F. (eds.) MICCAI 2015. LNCS, vol. 9351, pp. 234–241. Springer, Cham (2015). https://doi.org/10.1007/978-3-319-24574-4_28
26. Balakrishnan, G., Zhao, A., Sabuncu, M.R., Guttag, J., Dalca, A.V.: VoxelMorph: a learning framework for deformable medical image registration. IEEE Trans. Med. Imaging **38**(8), 1788–1800 (2019)
27. Fushiki, T.: Estimation of prediction error by using k-fold cross-validation. Stat. Comput. **21**, 137–146 (2011)
28. DeCost, B.L., Holm, E.A.: A computer vision approach for automated analysis and classification of microstructural image data. Comput. Mater. Sci. **110**, 126–133 (2015)
29. Jacobacci, F., et al.: Improving spatial normalization of brain diffusion MRI to measure longitudinal changes of tissue microstructure in the cortex and white matter. J. Magn. Reson. Imaging **52**(3), 766–775 (2020)
30. Klein, S., Staring, M., Murphy, K., Viergever, M.A., Pluim, J.P.: Elastix: a toolbox for intensity-based medical image registration. IEEE Trans. Med. Imaging **29**(1), 196–205 (2009)
31. ADNI Homepage. https://adni.loni.usc.edu/. Accessed 2 Aug 2023
32. Somerville, L.H., et al.: The lifespan human connectome project in development: a large-scale study of brain connectivity development in 5–21 year olds. Neuroimage **183**, 456–468 (2018)
33. Andersson, J.L., Graham, M.S., Zsoldos, E., Sotiropoulos, S.N.: Incorporating outlier detection and replacement into a non-parametric framework for movement and distortion correction of diffusion mr images. Neuroimage **141**, 556–572 (2016)
34. Andersson, J.L., Graham, M.S., Drobnjak, I., Zhang, H., Campbell, J.: Susceptibility-induced distortion that varies due to motion: correction in diffusion MR without acquiring additional data. Neuroimage **171**, 277–295 (2018)

Automatic Fast and Reliable Recognition of a Small Brain White Matter Bundle

John Kruper$^{(\boxtimes)}$ (ID) and Ariel Rokem (ID)

University of Washington, Seattle, USA
{jk232,arokem}@uw.edu

Abstract. Large diffusion MRI datasets provide the statistical power necessary to model complex effects in the white matter. They also motivate the need for fully automatic algorithms for finding white matter bundles. One popular algorithm, Automated Fiber Quantification (AFQ), has been shown to be reliable for analyzing a suite of large bundles. Here, we demonstrate that this approach can be extended to a relatively small white matter bundle, the optic tract. We develop an automated method that finds a portion of this bundle automatically. We compare the automatically found optic tract to optic tracts previously found using an algorithm that requires expert intervention, and find high degree of overlap. While previous methods work well in high-quality data, we demonstrate that the novel method proposed here generalizes broadly in subjects from three different datasets with differing data quality and a broad age range. Finally, we describe how this approach could be easily extended to other small bundles.

Keywords: Computational Diffusion MRI · Retinogeniculate Pathway · Cranial Nerve · Tractography · UK Biobank · Human Connectome Project · Healthy Brain Network

1 Introduction

There are several methods for recognizing known white matter bundles via tractography. However, these existing methods tend to focus on large anatomical structures that extend over large distances, such as the superior longitudinal fasciculus or corticospinal tract. Smaller bundles (such as the acoustic radiations, the cranial nerves, or small cortico-cortical connections between nearby cortical areas) are of interest to researchers, but automated tools to delineate are limited. Methods to find these small white matter bundles instead rely on manual expert annotation to some degree. And while these methods work very well in small samples, large population neuroimaging datasets, such as the ABCD study [1] and the UK Biobank [2] are too large to be manually processed and the field therefore needs robust automated recognition methods to study these smaller structures at the scale of these samples.

In this study, we introduce a fully automated method for recognizing and characterizing small brain white matter bundles based on the "Automated Fiber

M. Karaman et al. (Eds.): CDMRI 2023, LNCS 14328, pp. 70–79, 2023.
https://doi.org/10.1007/978-3-031-47292-3_7

Quantification" (AFQ) approach [3]. AFQ is a tractography-based approach to detect bundles by utilizing anatomical landmarks referred to as regions of interest (ROIs). These ROIs are typically placed in the deep white matter along the trajectory of each bundle and are defined in a template space. The ROIs are registered with each individual scan and used to find a particular bundle. Then, diffusion properties are quantified along the trajectory of that bundle [4]. This approach is fully automatic and has been shown to be reliable in large bundles [5]. However, it can be difficult to fully automate this approach in smaller bundles, where ROIs are not yet defined, and when defined can be either too small and difficult to register, or too large and contain more than just the intended bundle. To overcome these challenges, we propose (1) to automatically define ROIs based on a researcher-defined trajectory and (2) a series of refinement steps that effectively isolate the intended small bundles after the initial ROI identification is used. To demonstrate this technique, we quantify the tissue properties of the optic tract in three separate datasets, which together span a large range of ages and were collected with different instruments and a range of different acquisition sequences. The optic tract is the portion of the retinogeniculate visual pathway (RGVP) from the optic chiasm (where nerves coming from each eye cross) to the lateral geniculate nucleus (LGN) in the thalamus, and is an example of a small and hard-to-delinate bundle.

1.1 Related Work

Previous work focused on the retinogeniculate pathway often relies on ROIs that are hand-drawn in each subject. For example, using an automatically identified optic chiasm and manually drawn LGN ROIs, it has been shown that the optic tract tissue properties are correlated with primary visual cortex size [6] and are significantly different in patients with glaucoma [7]. In He et al. [8], an expert neurosurgeon drew five regions of interest on a map derived from the diffusion data for each of the 57 subjects included in the study. They employed a second expert neurosurgeon to do the same task to assess the reliability between experts, which was found to range between a weighted Dice of 0.74 and 0.81, depending on the tractography algorithm that was used. Hand drawn ROIs have merit, but this technique cannot be scaled to thousands of subjects easily, and subject to individual rater error, because although inter-rater reliability can be rather high, it is not perfect. Another approach is to use ROIs automatically generated from other modalities such as T1w images [9]. Such an approach is indeed fully automatic and has been shown to be useful in recognizing smaller bundles [10]. However, this previous approach focuses on where streamlines start and end, while the present work extends and improves on this approach by focusing on ROIs that are in the core white matter and the shape of the streamlines. These are complimentary techniques, and the best approach may depend on the bundles or the data.

2   Methods

2.1   Data

The UK Biobank [2] is an openly available dataset which has collected 45,000 diffusion MRI scans to date. Here, we processed 90 randomly selected subjects. We also processed data from 10 scans from the Human Connectome Project (HCP) [11]. Finally, we processed three randomly chosen subjects from the Healthy Brain Network (HBN), which is a landmark pediatric study that will eventually include over 5,000 diffusion MRI scans [12,13].

2.2   Bundle Recognition

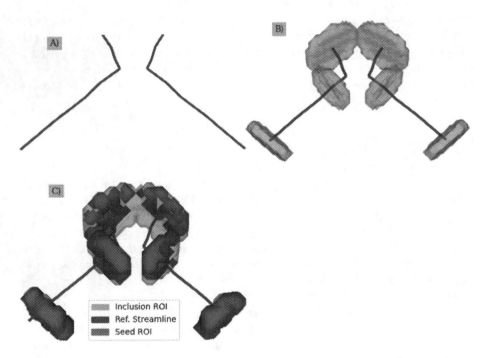

Fig. 1. A Hand drawn trace of the anterior optic tract and optic chiasm using a Montreal Neurological Institute (MNI) template [14] in blue. **B** Inclusion ROIs in MNI space in green, generated from the traces in A. **C** Inclusion ROIs and reference streamlines moved into subject space, with seed ROIs for tractography shown in red. (Color figure online)

We used the Montreal Neurological Institute template [14] to draw two reference streamlines over the anterior optic tract and optic chiasm, shown in Fig. 1A. These reference streamlines do not extend the entire length of the optic tract, because we only require successful delineation of the optic tract where it is most isolated. This makes recognition easier and reduces concerns from crossing fibers. Next, we generated 3 disks along and perpendicular to each of the reference streamlines to use as inclusion ROIs, shown in Fig. 1B. For each subject, we used generalized q-sampling imaging (GQI) [15] to estimate spherical harmonics from the diffusion data. We used these spherical harmonics to calculate anisotropic power (AP) [16], an index of anisotropy which is robust to noise, model independent, and particularly sensitive in low-anisotropy regions. Finally, in Fig. 1C, the reference streamline and inclusion ROIs are moved into subject space, and then seed ROIs are made for tractography where the inclusion ROIs have sufficiently high AP. Here, we show a typical subject from the HCP dataset.

To generate the tractography in the UK Biobank, we placed 8,000 seeds per voxel in the small seed ROI regions, which are around 150 voxels in size (example shown in Fig. 1C). In the higher resolution HCP dataset, we used only 216 seeds per voxel to generate a similar number of streamlines. We used a GPU-accelerated implementation of residual bootstrap tractography [17] in the UK Biobank and HCP datasets. In HBN, residual bootstrap tractography failed to track the optic tract, so we used probabilistic constrained spherical deconvolution [18] instead, with the same seeding strategy as in the UK Biobank.

After generating a tractography solution, we used 4 filtering steps to recognize each bundle. First, we filtered using inclusion ROIs, retaining streamlines only if they passed through all of these ROIs. Results shown in a typical HCP subject in Fig. 2A. For all further analysis, we only considered the parts of the streamline between the outside two inclusion ROIs. Next, we filtered streamlines by their curvature. To do this, we used a reference streamline, which describes the modal trajectory of the bundle of interest. The reference streamline is defined in the template space. This step needs to be done only once for each bundle of interest and is reused across datasets. Here, we drew the reference streamline, shown in Fig. 1C, but this reference streamline could also be selected from an already-segmented bundle, or based on an atlas. We resampled each streamline selected in the previous step to have the same number of points as the reference streamline, in this case, 12. We calculated the 3D gradient between each point in the reference and in each of the subject streamlines. We then calculated the average angle between the gradients of the reference streamline and each subject streamline. Only streamlines with an average angle of less than $20°C$ were accepted, shown in Fig. 2B. Next, we used QuickBundles [19] with a threshold of 6mm to cluster the streamlines selected in previous steps, selecting only streamlines belonging to the largest cluster, shown in Fig. 2C. This is particularly useful in cases where tractography solution trajectories are not unimodally distributed. Finally, we selected streamlines that pass a "cleaning" process commonly used in previous work [3]: an iterative procedure was used where, in each round, streamlines are removed if they are outliers in terms of both their length and their Mahalanobis distance from the average streamline trajectory. We did this for 5 rounds with a distance threshold of 2σ and a length threshold of 3σ, with results shown in Fig. 2D.

All of the subject-specific steps are handled automatically using pyAFQ [5] which relies heavily on DIPY [20]. We provide a reference configuration file for

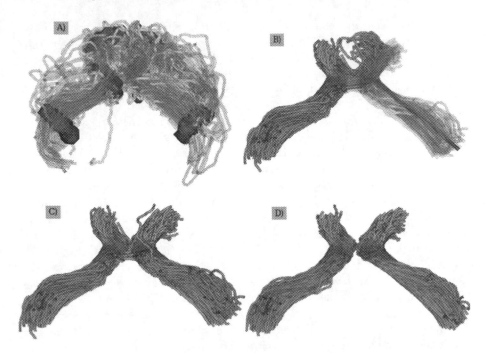

Fig. 2. A Streamlines selected by inclusion ROIs, with long streamlines removed for visual clarity (this is the only panel with streamlines removed for visual clarity). Inclusion ROIs are in red. **B** Streamlines remaining after filtering by curvature. The reference streamline is in red and only shown for the right optic tract, though each hemisphere has its own reference streamline. **C** Streamlines remaining after filtering by Quick-Bundles. **D** Streamlines remaining after Mahalanobis distance cleaning. (Color figure online)

pyAFQ and here are the inclusion ROIs and reference streamlines we used. To run this pipeline for a different small bundle, simply modify the configuration file to point to your own custom ROIs and/or reference streamlines. To generate custom ROIs from reference streamlines use this script.

To measure the success of this algorithm in finding the optic tract, we compared the optic tracts recognized by our algorithm with those found in He et al. [8]. In this previous work, ROIs were manually drawn in each subject by experts to find the entire RGVP. We quantify overlap using a one-sided weighted dice coefficient (owDSC) [21]:

$$\frac{\sum_{v'} W_{i,v'}}{\sum_{v} W_{i,v}}$$

where v is all voxels, v' is only voxels where the two tracts intersect, and W is the fraction of streamlines in a given tract passing through that voxel.

Finally, we ran RecoBundles [22] in all three datasets to find the optic tracts using the same tractographies that we used for our algorithm. We used a model cluster threshold of 1.25, a reduction threshold of 25 mm, and a pruning threshold of 12. We cleaned the results of recobundles using a similar procedure to our algorithm. We started by using Quickbundles [19] with a threshold of 6 mm and then used Mahalanobis distance cleaning with a distance threshold of 3σ and a length threshold of 3σ.

3   Results

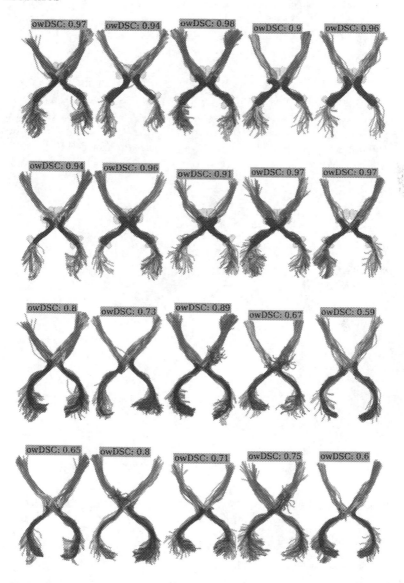

Fig. 3. Visualization of optic tracts found by our algorithm or Recobundles (in blue) and the RGVP found by He et al. [8] (in red) with the one-sided weighted dice coefficient (owDSC) shown in each panel. The top two rows show our algorithm, while the bottom two rows show the results of Recobundles. Waypoint ROIs are shown in green for our algorithm. (Color figure online)

The procedure that we developed can accurately delineate the optic tract in all individuals that we used in this study. For example, a visual examination of the 10 example HCP subjects shown in Fig. 3 demonstrates that it reliably recognized the same portion of the optic tract and optic chiasm, despite individual variability in the shape and size of these structures. For these 10 subjects, the average owDSC is 95% between our automatically recognized optic tracts and the RGVP recognized with the help of experts in [8] (min: 0.9, max: 0.98). For the entire process, including fitting GQI, registration to the template, tractography, bundle recognition, and visualization, each subject was processed in approximately 30 min.

In HCP, RecoBundles was also able to successfully find the optic tract in all individuals. Average owDSC is lower for RecoBundles, however, visual inspection validates that they follow the correct trajectory, and differences in owDSC are likely driven by differences at the endpoints. Additionally, RecoBundles recognizes the entire optic tract, from the thalamus to the optic chiasm, while our algorithm focuses only on the anterior portion, which is much easier to delineate (Fig. 4).

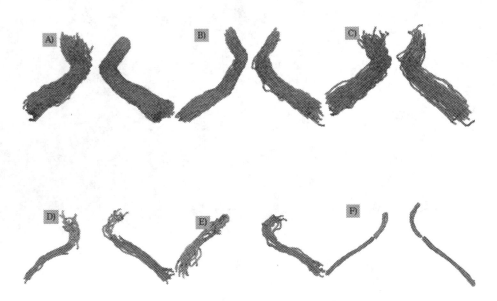

Fig. 4. Visualization of optic tracts found in HBN (A–C) and the UK Biobank (D–F). UK Biobank results show division into optic tract and optic chiasm sections.

To test the robustness of our algorithm across different datasets, with a range of subject characteristics and acquisition protocols, we also processed subjects from HBN and the UK Biobank. We processed three randomly selected subjects from HBN, shown in Fig. 3A–C. Of the 90 subjects in the UK Biobank we selected for processing, we automatically found the optic tract in all 90. Three randomly selected subjects are shown in Fig. 3D–F. Here, we also show a division between the optic tract and optic chiasm sections that may be useful for further analysis.

For these 6 subjects from these 2 datasets, RecoBundles was unable to delineate the optic tract. This highlights the crucial advantage of our algorithm - robustness across datasets of varying quality.

4   Discussion

The main contribution of this work is in developing an automated procedure to find small white matter pathways. We demonstrated the efficiency and reliability of this algorithm in detecting a portion of the retinogeniculate visual pathway automatically. We benchmarked the accuracy of the method with respect to a previous method that required expert intervention and found high overlap. In contrast to a previous state-of-the-art method, the RecoBundles algorithm [22], we demonstrated that the procedure generalizes well across different quality of data and a range of ages, with results in three separate datasets. For computational efficiency, we rely on an open-source GPU-accelerated tractography [17] and on other open-source software components (i.e., pyAFQ [5] and DIPY [20]), making it both fast and accessible.

This algorithm consists of a combination of a novel combination of pre-existing methods and the introduction of novel methods. The pre-existing methods include filtering based on inclusion ROIs as well as cleaning using Quick-Bundles [19] and Mahalanobis distance. Inclusion ROIs limit the curvature comparison to the core of the white matter. Cleaning by Mahalanobis distance may encounter issues when the distributions of streamline lengths and distances are not uni-modal, but this problem is effectively addressed by QuickBundles.

The two novel steps include curvature filtering and automated generation of waypoint ROIs. In contrast to RecoBundles, we found that curvature filtering works well in a range of datasets, including HBN and UK Biobank. Also, a single reference curve is easier to draw than an entire reference bundle. However, in the dataset where RecoBundles was successful, it was able to recognize the entire optic tract, and capture some of its fanning behavior near the thalamus. This highlights a limitation of our approach. Many bundles are known to fan out, particularly near the gray-white matter interface. In these cases, a single reference streamline is not sufficient to recognize all of the fanning streamlines that would be generated. For any analysis focusing on micro-structural tissue properties in the core white matter, this limitation is less consequential, as the curvature analysis is only be done in the core white matter (between inclusion ROIs).

There are additional limitations to this algorithm. We only recognize the portion of the optic tract which is easiest to track and recognize. This limitation is driven by the underlying data, and means that we use the portion of the optic tract with the least number of crossing fibers, which are potential confounders in further analysis. Another limitation is in the optic chiasm, where we do not attempt to recognize decussating fibers, which are more difficult to track, particularly in clinical-quality datasets. Additionally, the method described here is not intended to discover novel connections.

Nevertheless, this algorithm has great potential to be adapted to other small fibers that are already known from other anatomical methods. One need only draw a single reference streamline in MNI space. Then inclusion ROIs are drawn automatically as perpendicular disks along the reference streamline. This is more convenient than drawing inclusion ROIs manually, and it standardizes the nature of inclusion ROIs across different bundles.

Acknowledgements. Data were provided in part by the Human Connectome Project, WU-Minn Consortium (Principal Investigators: David Van Essen and Kamil Ugurbil; 1U54MH091657) funded by the 16 NIH Institutes and Centers that support the NIH Blueprint for Neuroscience Research; and by the McDonnell Center for Systems Neuroscience at Washington University. This research has been conducted using the UK Biobank Resource under Application Number 28541. Work on this project was funded by NSF grant 1934292, and NIH grants RF1 MH121868, NIA/NIH U19AG066567. We would like to thank the Child Mind Institute Biobank for access to the Healthy Brain Network dataset. We are also grateful to Fan Zhang and Lauren O'Donnell for sharing data from [8].

References

1. Jernigan, T.L., Brown, S.A., Dowling, G.J.: The adolescent brain cognitive development study. J. Res. Adolesc. **28**(1), 154–156 (2018)
2. Sudlow, C.: UK biobank: an open access resource for identifying the causes of a wide range of complex diseases of middle and old age. PLoS Med. **12**(3), e1001779 (2015)
3. Yeatman, J.D., Dougherty, R.F., Myall, N.J., Wandell, B.A., Feldman, H.M.: Tract profiles of white matter properties: automating fiber-tract quantification. PloS One **7**(11), e49790 (2012)
4. Jones, D.K., Travis, A.R., Eden, G., Pierpaoli, C., Basser, P.J.: PASTA: pointwise assessment of streamline tractography attributes. Magn. Reson. Med. **53**(6), 1462–1467 (2005)
5. Kruper, J., et al.: Evaluating the reliability of human brain white matter tractometry. Apert. Neuro **1**(1) (2021)
6. Miyata, T., Benson, N.C., Winawer, J., Takemura, H.: Structural covariance and heritability of the optic tract and primary visual cortex in living human brains. J. Neurosci. **42**(35), 6761–6769 (2022)
7. Ogawa, S., et al.: Multi-contrast magnetic resonance imaging of visual white matter pathways in patients with glaucoma. Invest. Ophthalmol. Vis. Sci. **63**(2), 29 (2022)
8. He, J., et al.: Comparison of multiple tractography methods for reconstruction of the retinogeniculate visual pathway using diffusion MRI. Hum. Brain Mapp. **42**(12), 3887–3904 (2021)
9. Lerma-Usabiaga, G., Liu, M., Paz-Alonso, P.M., Wandell, B.A.: Reproducible tract profiles 2 (RTP2) suite, from diffusion MRI acquisition to clinical practice and research. Sci. Rep. **13**(1), 6010 (2023)

10. Liu, M., Lerma-Usabiaga, G., Clascá, F., Paz-Alonso, P.M.: Reproducible proto-
    col to obtain and measure first-order relay human thalamic white-matter tracts.
    Neuroimage **262**, 119558 (2022)
11. Van Essen, D. C., Smith, S. M., Barch, D. M., Behrens, T. E., Yacoub, E., Ugurbil,
    K., Wu-Minn HCP Consortium: The WU-Minn human connectome project: an
    overview. Neuroimage **80**, 62–79 (2013)
12. Alexander, L.M., et al.: An open resource for transdiagnostic research in pediatric
    mental health and learning disorders. Sci. Data **4**, 170181 (2017)
13. Richie-Halford, A., et al.: An analysis-ready and quality controlled resource for
    pediatric brain white-matter research. Sci. Data **9**(1), 616 (2022)
14. Fonov, V., Evans, A.C., Botteron, K., Almli, C.R., McKinstry, R.C., Collins, D.L.,
    Brain Development Cooperative Group: Unbiased average age-appropriate atlases
    for pediatric studies. Neuroimage **54**(1), 313–327 (2011)
15. Yeh, F.C., Wedeen, V.J., Tseng, W.Y.I.: Generalized q-sampling imaging. IEEE
    Trans. Med. Imaging **29**(9), 1626–1635 (2010)
16. Dell'Acqua, F., Lacerda, L., Catani, M., Simmons, A.: Anisotropic power maps: a
    diffusion contrast to reveal low anisotropy tissues from HARDI data. Proc. Intl.
    Soc. Mag. Reson. Med. **22**, 29960–29967 (2014)
17. Rokem, A., et al. GPU-accelerated diffusion MRI tractography in DIPY. Proc.
    Intl. Soc. Mag. Reson. Med. **21** (2021)
18. Tournier, J.D., Calamante, F., Connelly, A.: Robust determination of the fibre ori-
    entation distribution in diffusion MRI: non-negativity constrained super-resolved
    spherical deconvolution. Neuroimage **35**(4), 1459–1472 (2007)
19. Garyfallidis, E., Brett, M., Correia, M.M., Williams, G.B., Nimmo-Smith, I.:
    Quickbundles, a method for tractography simplification. Front. Neurosci. **6**, 175
    (2012)
20. Garyfallidis, E., et al.: Dipy, a library for the analysis of diffusion MRI data. Front.
    Neuroinf. **8**, 8 (2014)
21. Cousineau, M., et al.: A test-retest study on parkinson's PPMI dataset yields
    statistically significant white matter fascicles. NeuroImage Clin. **16**, 222–233 (2017)
22. Garyfallidis, E.: Recognition of white matter bundles using local and global
    streamline-based registration and clustering. Neuroimage **170**, 283–295 (2018)

Self Supervised Denoising Diffusion Probabilistic Models for Abdominal DW-MRI

Serge Vasylechko[(⊠)], Onur Afacan, and Sila Kurugol

QUIN Lab, Department of Radiology, Boston Children's Hospital, Harvard Medical School, Boston, USA
serge.vasylechko@childrens.harvard.edu

Abstract. Quantitative diffusion weighted MRI in the abdomen provides important markers of disease, however significant limitations exist for its accurate computation. One such limitation is the low signal-to-noise ratio, particularly at high diffusion b-values. To address this, multiple diffusion directional images can be collected at each b-value and geometrically averaged, which invariably leads to longer scan time, blurring due to motion and other artifacts. We propose a novel parameter estimation technique based on self supervised diffusion denoising probabilistic model that can effectively denoise diffusion weighted images and work on single diffusion gradient direction images. Our source code is made available at https://github.com/quin-med-harvard-edu/ssDDPM

Keywords: denoising · diffusion probabilistic models · quantitative mapping · abdominal MRI

1 Introduction

Abdominal diffusion-weighted magnetic resonance imaging (DW-MRI) has been shown to improve disease assessment and especially the assessment of tumor response after oncological treatment [17]. Evidence suggests that DW-MRI based apparent diffusion coefficient (ADC) values are helpful in early detection, characterization and in follow-up of various tumor types [1]. For example, ADC values frequently occur before changes in tumor size [4]. To compute ADC, DW-MRI data at different diffusion strengths (b-values) are fitted to an exponential decay model. However, signal to noise ratio (SNR) is especially low at high b-values, which reduces robustness and reproducibility of ADC estimates. To address this, for each b-value, multiple images are typically acquired with different diffusion gradient directions and are then geometrically averaged to improve SNR. In this instance, diffusion directional information is assumed to be isotropic for the purpose of ADC fitting. Such acquisition strategy comes at the expense of linearly increased scan time, increased cost and more discomfort to the patient [18]. Furthermore, in abdominal imaging in particular, such averaging techniques are

© The Author(s), under exclusive license to Springer Nature Switzerland AG 2023
M. Karaman et al. (Eds.): CDMRI 2023, LNCS 14328, pp. 80–91, 2023.
https://doi.org/10.1007/978-3-031-47292-3_8

often detrimental to the fine details of the underlying anatomy due to inherent and unavoidable respiratory motion. Significant misalignments are typical between voxels as the individually acquired images are often shifted in space. The goal of this study is therefore to improve ADC estimation for low SNR DW-MRI data that consists of a single image at each b-value at test time.

Many denoising techniques that have originally been developed for natural images can also be applied to DWI-MRI, such as non-local means filters (NLM), block-matching (BM3D), and K-singular value decomposition (K-SVD) [2,3,6]. Other MRI reconstruction methods that make use of prior knowledge, such as sparseness and low rank, can also achieve good denoising effects [8].

A different set of restoration methods exploit the additional information present in multiple images of different diffusion directions of the same b-value, such as the Marchenko-Pastur principal component analysis (MPPCA), Patch2Self and DDM2 [7,16,19]. Partial redundancy can be exploited for denoising based on the statistical independence of noise. The core requirement of these methods is the availability of multiple diffusion direction images at each b-value.

Supervised training can also be used on pairs of high and low SNR images to train an effective denoising network. 1D (voxelwise), 2D and 3D CNN based networks have been introduced to this purpose [5,10,14]. The downside of such approaches is the need to train a separate denoising network for each b-value, and the need to collect relevant motion free data.

More recently, self supervised CNN based method have been introduced that estimates bi-exponential decay parameters from low SNR DW-MRI data directly, and had shown to improve robustness under the constraints of noisy diffusion signal [15]. Parameters are predicted by 2D Unet, followed by a signal estimation with the forward model on the GPU, from which the loss is then computed with respect to the original input signal for self supervision.

In this work, our goal is to eliminate the need to acquire multiple diffusion directions at each b-value while estimating robust ADC coefficients. We formulate a novel self supervised based training approach that combines a denoising diffusion probabilistic model (DDPM) and self supervised parameter estimation that is tailored to multi-b-value DW-MRI images with ADC fitting constraints.

2 Method

2.1 Denoising Diffusion Model

We are given a dataset of multiple diffusion gradient direction images at multiple b-values, denoted as $D = \{x_{b,r,i}\}$, where b refers to the b-value, r refers to the diffusion gradient direction (effectively treated as a repetition in ADC estimation) and i is a subject sample drawn from a distribution of all possible DW-MRI images. Consider true source samples y, such that $x_{b,r,i} = y_{b,r,i} + \epsilon$, where ϵ is some noise distribution. By treating r as a repetition in ADC setting, we can construct an approximation $\hat{y}_{b,i}$ by averaging our samples across multiple diffusion gradient directions $\sum_{r=1}^{R}(x_{b,r,i})/R$. However, in abdominal imaging the repetitions, $x_{b,r,i}$, are not anatomically aligned on a voxel-to-voxel basis due to unavoidable respiratory motion, which cannot be easily remedied with motion correction algorithms. Hence, we could not train a denoising network in supervised manner on pairs of $\{\hat{y}_{b,i}, x_{b,r,i}\}$ for some arbitrarily chosen r. Instead, during training, we propose to add the correct amount of synthetically generated noise ϵ, sampled from some distribution, e.g. Gaussian $\epsilon \sim \mathcal{N}(\mu, \sigma^2)$, to $\hat{y}_{b,i}$, without using $x_{b,r,i}$ entirely at training time. Our task is to model correct $\{\mu, \sigma\}$ of the noise distribution, such that $\{\hat{y}_{b,i}, x_{b,r,i}\}$ relationship is upheld in terms ϵ, as we would like to use $x_{b,r,i}$ samples as input to our trained network at test time only.

A DDPM [9] can be thought of as a generalized denoiser. The sum of Gaussian independent random variables is a Gaussian random variable. Thence, the forward noise corruption process can be modelled as a Markov Chain, where noise is added incrementally to the image over and over again. We can control the penultimate amount of noise added, $\{\mu, \sigma\}$, by controlling only the number of timesteps in a Markov Chain, $t = T$. To do this we compare SNR between a noise corrupted samples $\hat{y}_{b,i}$ for each $t = \{0..T\}$, and actual samples of $x_{b,r,i}$, and find the optimal $\hat{T}$ to use at test time.

Formally, clean data distribution is represented by $q(y_0)$, and given a data sample $\hat{y}_{b,i} \sim q(y_0)$, a forward noise corruption process p produces latent y_t through y_T with a noise schedule $\beta_1, ..., \beta_T$. During training, we aim to recover an estimate of y_{t-1}, given by $\hat{y}_{t-1}$, for a randomly sampled t from a uniform distribution for each minibatch. The inverse process is parametrized by a learnable network with weights f_θ via gradient descent:

$$\nabla_\theta \left\| f_\theta \left(\sqrt{1 - \beta_t} y_0 + \beta_t \epsilon, t \right) - \epsilon \right\|_p^p \tag{1}$$

Once the network has been trained, the reverse process will require inference over the entire sequence of $\{T..0\}$ backwards as a Markov Chain.

$$\hat{y}_{t-1} = \frac{1}{\sqrt{1 - \beta_t}} \left(\hat{y}_t - \frac{\beta_t}{\sqrt{1 - \gamma_t}} f_\theta (\hat{y}_t, t) \right) + \sqrt{\beta_t} \epsilon \tag{2}$$

However, we would like to guide the denoiser f_θ to output noise estimates ϵ in accordance to the best possible ADC fit for each backwards $t - 1$ step in the Markov chain. Hence, we concurrently train an additional network that takes the output of f_θ, computes $\hat{y}_{t-1}$, and feeds the result to a parameter estimation network.

2.2   Self Supervised Parameter Estimation

ADC diffusion model is described by $S(b) = S_0 e^{-bD}$, where b is the vector of diffusion weightings (b-values), S_0 is the non-diffusion dependent MR signal, D is the apparent diffusion coefficient. The exponential parameter fit can be reformulated in a closed loop form as a singular value decomposition problem [11] to solve the inverse problem by estimating the two parameters of the ADC model from $\hat{y}_{t-1}$. The closed loop form allows us to embed this directly into the Pytorch framework, such as that the parameters can be computed all at once for all the images in a batch in a single step. Once the parameters are estimated, we feed them to the loss function, which is formed of the ADC diffusion forward model that takes the network predicted S_0 and D parameters and computes the estimate of the original input L2 norm difference with y_{t-1}, which is the image produced by the forward noise corruption process p. No ground truth parameters are needed for training such network. An overview of the proposed training is given in Fig. 1.

2.3   Implementation and Training

Noise scheduler p and the network architecture for DDPM follows the work of Ho et al. [9] and is implemented in Pytorch on top of Diffusers library [13]. 2D UNet architecture consisted of 4 blocks with 2 layers per block, with 64, 128, 128, 256 channels respectively. 2 out of 4 block transitions contain spatial self-attention layers. Input images are resized (downsampled) to 192×192 grayscale. Input consists of 3 channels that correspond to vector of b-values (b=50,400,800 in our study) and the output of DDPM is 1 channel that corresponds to predicted ϵ with fixed variance [12]. Linear noise scheduler parameters are set to denoise $\hat{y}_{b,i}$ such as to correspond to low SNR input, $x_{b,r,i}$, which are β_1, β_T, T , given by 1e-7, 2e-6, and 250 respectively. The second part of the architecture, namely the SVD based ADC parameter estimation, does not have any training parameters, with all computations consisting of matrix manipulations loaded on the GPU. The input to the ADC estimation are 3 channels, that corresponds to $\hat{y}_{(t-1)}$, and output is 2 channels that corresponds to D and $S0$ estimates. The model was trained for 200 epochs with batches of 8 images on 60,000+ individual 2D slices.

2.4   Data

We performed an IRB approved and HIPAA compliant study. Training data consisted of abdominal DW-MRI data retrospectively acquired from a 3T Siemens scanner according to the clinical protocol from 105 subjects, with b-values = 50, 400, 800 s/mm2, and 6 diffusion gradient directions. A free-breathing single-shot EPI sequence was based on: TR/TE=5200/78 ms; FOV= 380×310 mm; in-plane resolution=1.5×1.5 mm2; slice thickness = 4 mm; BW = 2442 Hz/px. Test data was acquired on 5 volunteer subjects using the same clinical protocol. Additionally, a special radio frequency transmitter, termed PilotTone, was placed on the patient table next to the hip of the subject, which allowed for respiratory motion trajectory to be gathered for this test data. The key aim was to further reduce residual effects of motion between b-values of individual NEX1 images for more accurate evaluation. The data was binned and aligned according to respiratory phases.

2.5   Experiments

Our goal is to evaluate if the proposed method could reasonably improve the ADC fit over the low SNR images (NEX1, equivalent to a single diffusion gradient direction images), $x_{b,r,i}$, and yield data that is qualitatively similar to high SNR images (NEX6, equivalent to 6 diffusion gradient directions that were geometrically averaged), $\hat{y}_{b,i}$. We evaluate the accuracy of denoising on ADC parameter fitting error by looking at the residuals between the fitted ADC maps on the NEX6 signal, and its noise corrupted variant. We suppose that reduction in noise that is guided to be estimated in the direction of the best ADC fitting, should yield to better accuracy with smaller residuals, which we measure in terms of MSE. We evaluate the proposed method in clinically acquired data. For comparison, we test our approach against the NLM algorithm, as well as the conventional DDPM algorithm.

To produce realistic noise corruption for NEX6 data, SNR was measured in multiple single repetition scans (NEX1) and compared to corresponding NEX6 images at each timestep t of the forward diffusion process. An assumption was made that any residual motion effects are negligible for SNR calculation with sufficiently large segmentation masks over the abdomen and background area. Noise schedule parameters were adjusted to yield a sufficient number of timesteps to ensure smooth diffusion process for the given β_0 and β_T such that the SNR would reach similar levels as NEX1 images at the final timestep T.

To evaluate precision and accuracy of the denoised single diffusion gradient direction data, an additional experiment was conducted to acquire volunteer breath hold DWI data. A breath hold enabled us to geometrically average multiple diffusion gradient directions at each b-value, thereby constructing a high SNR 'reference' image (NEX3) without worrying about the effects of motion. Acquisition settings were the same as the clinical protocol, except for a three times longer scan time compared to NEX1 acquisition and a smaller coverage of only 10 slices through the middle section of the abdomen, to be able to fit the acquisition of all 3 bvalues and 3 gradient directions into a single breathold. ADC estimates from each of the 3 denoised single diffusion gradient direction images (NEX1) were compared to each other to calculate the coefficient of variation, as a measure of precision. Also ADC values from each NEX1 image was compared to the ADC values of the reference NEX3 image, as a measure of accuracy. Quantitative comparison metrics are reported in three regions of interest for clinical practical utility - liver (vessel free ROIs), kidney cortex and spleen.

3   Results

An example of the geometrically averaged B800 images from 6 diffusion gradient directions (NEX6) is shown in Fig. 2, together with the synthetically corrupted image that serves as input to the denoising algorithms. The noise corruption corresponds to the worst SNR level that was evaluated in our training data, and was obtained with the DDPM noise scheduler accordingly. NLM denoising leads to smoothing of certain edges in the image, as can be seen in the top right corner in Fig. 2. DDPM denoising and the proposed self supervised DDPM (ssDDPM) have qualitatively better adherence to the original NEX6 image. Notably, the dorsal parts of the image in the ssDDPM appear much more coherent.

Figure 3 shows the ADC map derived from reference image (NEX3), as well as denoised versions of NEX1 image with NLM, DDPM and ssDDPM. The DDPM denoised ADC map has good structural appearance, yet the most coherent appearance is from ssDDPM denoised data. Note the smaller kidney cortex 'holes' in the image for improved structural details.

Table 1 shows the measure of accuracy on motion free data that was acquired via a breathhold. A quantitative comparison is made between high SNR reference data (NEX3) and denoised low SNR data (NEX1). Mean and standard deviation of RMSE and PSNR are given across 3 repetitions in kidneys' cortex, spleen and liver. ssDDPM yields higher overall accuracy than comparison methods in all regions, with comparable variance. Table 2 provides a measure of precision on the same motion free data. Coefficient of variation (CoV) is estimated between 3 repetitions of the same scan (NEX1). ssDDPM yields better CoV in spleen and liver, but is surpassed by DDPM in the kidneys.

Table 3 evaluates RMSE and PSNR of ADC estimates over the volunteer test set for a synthetic noise corruption, where the ground truth NEX6 images were noise corrupted and then subsequently denoised. ssDDPM yields higher overall accuracy in this synthetic experiment than other methods except than in kidneys (similar to Table 2).

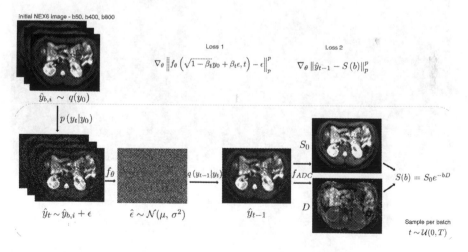

Fig. 1. An overview of the proposed method to train DDPM jointly with the self-supervised L2 loss from the ADC model predictions at each training step for each uniformly chosen corruption step t. Model based loss guides the noise prediction process towards minimizing the ADC fit.

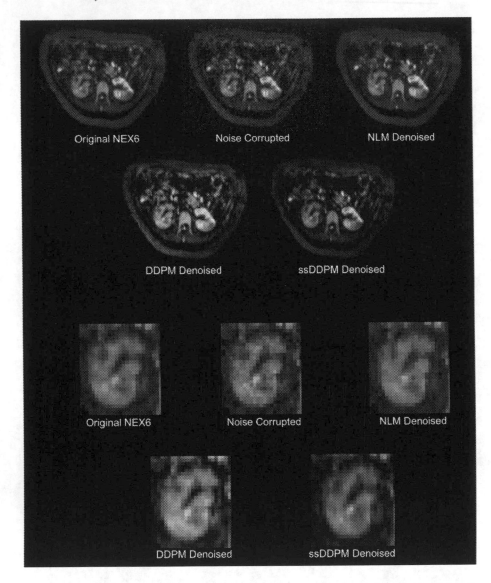

Fig. 2. Original NEX6 image, its noise corrupted variant, and the denoised results from NLM, DDPM and the proposed self supervised DDPM (ssDDPM). Top: whole abdomen view. Bottom: zoom on the left kidney. ssDDPM preserves structural details of the image better than the comparison methods.

Table 1. Measure of accuracy on motion free data (acquired via a breathhold) between high SNR reference data (3 geometrically averaged repetitions, NEX3) and denoised single scan data (NEX1). Mean and standard deviation of RMSE and PSNR is given across 3 repetitions in kidneys, spleen and liver.

	Kidneys		Spleen		Liver	
	RMSE (10e-5)	PSNR (dB)	RMSE (10e-5)	PSNR (dB)	RMSE (10e-5)	PSNR (dB)
NEX1	39.6 (7.5)	68.1 (1.6)	16.4 (6.0)	76.2 (3.4)	34.3 (6.1)	69.4 (1.6)
NLM	37.3 (6.4)	68.7 (1.5)	16.3 (6.8)	76.3 (3.5)	31.2 (6.4)	70.2 (1.9)
DDPM	39.4 (4.4)	68.1 (1.0)	14.9 (6.2)	77.2 (3.8)	30.7 (7.0)	70.3 (2.5)
ssDDPM	**33.6 (5.4)**	**69.5 (1.5)**	**13.9 (5.1)**	**77.6 (3.4)**	**27.0 (7.8)**	**71.6 (2.6)**

Table 2. Measure of precision on motion free data (acquired via a breathhold). Coefficient of variation (CoV) is calculated between 3 repetitions of the same scan (NEX1).

	CoV (%)		
	Kidneys	Spleen	Liver
NLM	2.2	4.7	2.6
DDPM	**1.4**	5.0	3.6
ssDDPM	2.1	**4.4**	**2.5**

Table 3. RMSE and PSNR between ADC estimates derived from the test set of clean NEX6 images and NEX6 images that were synthetically corrupted according to the noise schedule, and subsequently denoised with the competing methods. ssDDPM yields superior quantitative accuracy on this synthetic test set in comparison to NLM and DDPM in the liver and spleen. DDPM yields better estimates in the kidneys.

	Kidneys		Spleen		Liver	
	RMSE (10e-5)	PSNR (dB)	RMSE (10e-5)	PSNR (dB)	RMSE (10e-5)	PSNR (dB)
Noisy	6.3 (3.0)	84.6 (3.7)	8.8 (7.3)	83.5 (7.5)	14.2 (2.4)	77.0 (1.5)
NLM	4.2 (1.6)	88.0 (3.2)	5.8 (4.1)	86.4 (6.1)	8.3 (1.6)	81.7 (1.7)
DDPM	**3.3 (1.2)**	**90.0 (3.3)**	4.7 (2.9)	87.9 (5.4)	7.9 (1.5)	82.1 (1.6)
ssDDPM	4.3 (1.3)	87.6 (2.3)	**4.2 (1.4)**	**87.9 (2.9)**	**6.5 (2.2)**	**84.2 (3.4)**

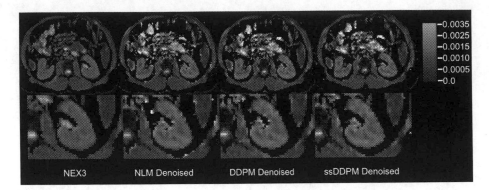

Fig. 3. An example of the ADC map obtained with reference image (NEX3), and ADC maps obtained after denoising low resolution NEX1 image with NLM, DDPM and ssDDPM methods respectively. ssDDPM yields the most structural coherent image, without oversmoothing effects.

4   Discussion and Conclusions

We proposed a new self supervised denoising and parameter estimation technique for abdominal DW-MRI which is based on DDPMs that can effectively denoise DW-MR images. The results showed that with the proposed method, the quality of the images and ADC maps were improved while achieving several folds reduced scan time compared to a baseline denoising method. The error in fitting the ADC model was also reduced when compared to the original NEX1 acquisition and the baseline denoising method. ssDDPM yields improved accuracy than competing methods in kidneys, spleen and liver in the breath hold experiment, as well as superior precision in spleen and liver. It has lower higher CoV than DDPM in the kidneys.

It is interesting to note that a higher overall accuracy was observed in the synthetic experiment, which comprised of a simple gaussian noise corruption model on NEX6 images. Conversely, when comparing geometrically averaged data in the breath hold experiment to individual NEX1 images the errors were substantially higher as noted by RMSE and PSNR. This is expected as the difference between NEX3 and NEX1 images cannot be described by pure gaussian noise effects, but may have small residual misalignment effects due to small motion and non isotropic diffusion gradient effects. This further highlights the crucial need to develop robust methods for parameter estimation that work on single gradient diffusion gradient direction data, as geometric averaging of multiple images does not necessarily yield more accurate estimates due to these residual effects. Interestingly, the relative overall CoV for the same breathhold experiment was relatively small in comparison (within single digit percentage points). Secondly, we also note that our method did not supercede standard DDPM in the kidneys in 2 out of 3 measures, which merits further investigation in the future for improved accuracy and precision.

The proposed method has several limitations that need to be addressed in future work. First, it assumes that the noise follows a Gaussian distribution, which is not always the case in practice. Single coil MRI reconstructed images are known to follow Rician noise distribution, however, we note that for multi-coil, EPI accelerated data the noise distribution becomes much more complex and may have long tails.

Another major drawback is that diffusion models are computationally expensive, as they require performing iterative steps at inference time to generate a sample. This can makes them impractical for use in real-time applications and significantly slower than other deep learning models. More efficient variants has been recently proposed and can be applied in future work.

Third, due to the Markovian chain, the diffusion denoising may introduce artifacts or distortions in the denoised images if not properly calibrated or if the SNR level in the original image is too low. Note that SNR levels may vary due to placement of coils. To counter-act this, a conditional diffusion model may be used.

Extended study may consider an investigation into the impact of the chosen ADC fitting algorithm on the results. In this work, we considered a linear least squares fitting that provides a closed form solution for ADC estimation [11], as it is an essential requirement for feeding the gradients back to the neural network at training time. This may be circumnavigated by using the ADC fitting in the markov chain at inference time only with an additional gradient descent term. In that scenario, a more accurate ADC fitting algorithm can be used such as nonlinear least squares fitting, or maximum likelihood based estimation that considers a more realistic noise model.

Additionally note that, unlike in brain DW-MRI, abdominal clinical DW-MRI protocols typically garner maximum b-values of $800–1000 \, s/mm^2$. Since this study focused on data acquired with a routine clinical protocol, which guaranteed both clinical relevancy and large quantities of data for training, the highest available b-value was $800 \, s/mm^2$. But the proposed method may potentially be effective to denoise images acquired at higher b-values.

Future experiments may also focus on evaluating the proposed method on adjunct quantitative fitting models for DW-MRI based data, such as intravoxel incoherent motion model (IVIM) and fractional anisotropy (FA). Unfortunately, both IVIM and FA require additional b-values and additional diffusion gradient directions respectively, for model parameter estimations, which was not possible with the current dataset used.

Acknowledgements. This work was supported partially by the National Institute of Diabetic and Digestive and Kidney Diseases (NIDDK), National Institute of Biomedical Imaging and Bioengineering (NIBIB), National Institute of Neurological Disorders and Stroke (NINDS) and National Library of Medicine (NLM) of the National Institutes of Health under award numbers R01DK125561, R21DK123569, R21EB029627, R01NS121657, R01LM013608, S10OD0250111 and by the grant number 2019056 from the United States-Israel Binational Science Foundation (BSF), and a pilot grant from National Multiple Sclerosis Society under Award Number PP-1905-34002.

References

1. Afaq, A., Andreou, A., Koh, D.: Diffusion-weighted magnetic resonance imaging for Tumour response assessment: why, when and how? Cancer Imag. **10**(1A), S179 (2010)
2. Aharon, M., Elad, M., Bruckstein, A.: K-SVD: an algorithm for designing overcomplete dictionaries for sparse representation. IEEE Trans. Signal Process. **54**(11), 4311–4322 (2006)
3. Bhujle, H.V., Vadavadagi, B.H.: NLM based magnetic resonance image denoising-a review. Biomed. Signal Process. Control **47**, 252–261 (2019)
4. Caro-Domínguez, P., Gupta, A.A., Chavhan, G.B.: Can diffusion-weighted imaging distinguish between benign and malignant pediatric liver tumors? Pediatr. Radiol. **48**, 85–93 (2018)
5. Cheng, H., et al.: Denoising diffusion weighted imaging data using convolutional neural networks. PLoS ONE **17**(9), e0274396 (2022)
6. Dabov, K., Foi, A., Katkovnik, V., Egiazarian, K.: Image denoising by sparse 3-d transform-domain collaborative filtering. IEEE Trans. Image Process. **16**(8), 2080–2095 (2007)
7. Fadnavis, S., Batson, J., Garyfallidis, E.: Patch2self: denoising diffusion MRI with self-supervised learning. Adv. Neural. Inf. Process. Syst. **33**, 16293–16303 (2020)
8. Haldar, J.P.: Low-rank modeling of local k-space neighborhoods (loraks) for constrained MRI. IEEE Trans. Med. Imag. **33**(3), 668–681 (2013)
9. Ho, J., Jain, A., Abbeel, P.: Denoising diffusion probabilistic models. Adv. Neural. Inf. Process. Syst. **33**, 6840–6851 (2020)
10. Jurek, J., et al.: Supervised denoising of diffusion-weighted magnetic resonance images using a convolutional neural network and transfer learning. Biocybern. Biomed. Eng. **43**(1), 206–232(2023)
11. Lupu, M., Todor, D.: A singular value decomposition based algorithm for multi-component exponential fitting of NMR relaxation signals. Chemom. Intell. Lab. Syst. **29**(1), 11–17 (1995)
12. Nichol, A.Q., Dhariwal, P.: Improved denoising diffusion probabilistic models. In: International Conference on Machine Learning, pp. 8162–8171. PMLR (2021)
13. von Platen, P., et al.: Diffusers: State-of-the-art diffusion models. https://github.com/huggingface/diffusers
14. Ran, M., et al.: Denoising of 3D magnetic resonance images using a residual encoder-decoder Wasserstein generative adversarial network. Med. Image Anal. **55**, 165–180 (2019)
15. Vasylechko, S.D., Warfield, S.K., Afacan, O., Kurugol, S.: Self-supervised IVIM DWI parameter estimation with a physics based forward model. Magn. Reson. Med. **87**(2), 904–914 (2022)
16. Veraart, J., Fieremans, E., Novikov, D.S.: Diffusion MRI noise mapping using random matrix theory. Magn. Reson. Med. **76**(5), 1582–1593 (2016)
17. Wang, Y.X.J., Huang, H., Zheng, C.J., Xiao, B.H., Chevallier, O., Wang, W.: Diffusion-weighted mri of the liver: challenges and some solutions for the quantification of apparent diffusion coefficient and intravoxel incoherent motion. Am. J. Nuclear Med. Molecular Imag. **11**(2), 107 (2021)

18. Winfield, J., et al.: Development of a diffusion-weighted MRI protocol for multi-centre abdominal imaging and evaluation of the effects of fasting on measurement of apparent diffusion coefficients (adcs) in healthy liver. Br. J. Radiol. **88**(1049), 20140717 (2015)
19. Xiang, T., Yurt, M., Syed, A.B., Setsompop, K., Chaudhari, A.: Ddm2: self-supervised diffusion mri denoising with generative diffusion models. arXiv preprint arXiv:2302.03018 (2023)

Voxlines: Streamline Transparency Through Voxelization and View-Dependent Line Orders

Besm Osman$^{(\boxtimes)}$ (iD), Mestiez Pereira (iD), Huub van de Wetering (iD), and Maxime Chamberland (iD)

Eindhoven University of Technology, Eindhoven, The Netherlands
b.osman@student.tue.nl, m.chamberland@tue.nl

Abstract. As tractography datasets continue to grow in size, there is a need for improved visualization methods that can capture structural patterns occurring in large tractography datasets. Transparency is an increasingly important aspect of finding these patterns in large datasets but is inaccessible to tractography due to performance limitations. In this paper, we propose a rendering method that achieves performant rendering of transparent streamlines, allowing for exploration of deeper brain structures interactively. The method achieves this through a novel approximate order-independent transparency method that utilizes voxelization and caching view-dependent line orders per voxel. We compare our transparency method with existing tractography visualization software in terms of performance and the ability to capture deeper structures in the dataset.

Keywords: Tractography · Visualization · Transparency · Streamlines

1   Introduction

Tractography datasets are growing larger, posing challenges for their visualization in terms of performance and usability. Various methods have been developed to address performance issues, such as removing colinear points or compressing the dataset [5]. Existing tractography visualization tools like MRtrix3 [9] and TrackVis [11] provide filtering options for visualizing different parts of the dataset. Recently, transparency-based methods have emerged to improve usability, such as applying varying transparency to each fiber based on orientation to highlight underlying tissue configurations [8]. Transparency can play an important role in tractography visualization by enabling exploration of deeper brain structures. Transparency has been used to better convey the spatial relationship between streamlines and surfaces that illustrate their anatomical context. [7] However, existing tractography visualization software, utilize subpar transparency methods due to performance constraints [4], limiting the benefits gained from transparency.

© The Author(s), under exclusive license to Springer Nature Switzerland AG 2023
M. Karaman et al. (Eds.): CDMRI 2023, LNCS 14328, pp. 92–103, 2023.
https://doi.org/10.1007/978-3-031-47292-3_9

1.1   Existing Transparency Methods

The basic approach to achieving transparency is by rendering the objects furthest from the screen first, which requires sorting every translucent object whenever the view changes. This method is suitable for visualization contexts with a limited number of transparent objects or a fixed viewing direction. However, this solution is not feasible for interactive tractography visualization. It is not possible to determine a consistent sorting order for entire streamlines, since different parts of a streamline can be at varying distances from the view. Therefore, sorting needs to be done at the level of individual line segments that make up the streamlines. The scale of the tractography dataset makes the computational cost of real-time sorting of line segments impractical. Hence, order-independent methods are required to achieve tractography visualizations with transparency. These techniques eliminate the need for ordering translucent objects. Various order-independent transparency techniques have been researched for 3D line sets, including depth peeling [1], multi-layer alpha blending [6], and raycast techniques [2]. In a recent comprehensive study [3], different order-independent transparency techniques for 3D linesets were compared, each showing different advantages and drawbacks. Some techniques require multiple render passes, resulting in decreased performance. Other techniques rely on complex render pipelines supported only by newer hardware or have high preprocessing times.

2   Method Overview

We propose a novel method for voxel-based streamline rendering that achieves fast, approximate order-independent transparency without relying on modern render pipelines and minimal preprocessing time. The first part of this method involves splitting the dataset into voxels, where each voxel stores streamline segments that pass through it as described in Sects. 2.1 and 2.2. Next, a mesh is generated for each voxel that connects the streamlines segments and renders the streamlines explained in Sect. 2.3. This mesh generation is performed once when reading the dataset. During rendering, we sort the voxels from back to front every frame, solving the most noticeable transparency issues. Furthermore, we extend this method by storing a set of precomputed line segment render orders per voxel and selecting the closest ordering based on the current view as described in Sect. 2.6 to improve the transparency accuracy within a voxel.

2.1   Voxelization

The algorithm takes a set of streamlines as input, where each streamline consists of a sequence of points. Our goal is to divide the dataset into voxels so that each voxel contains all line segments passing through it. In this paper, we define a voxel as a 3D cube with a fixed position and size. The size of the voxels used in this method differs from voxel sizes used in MRI scans or tractography algorithms. The voxels used in our method are an order of magnitude larger

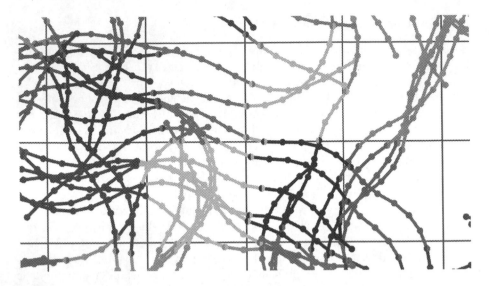

Fig. 1. This figure shows a 2D representation of the voxel meshes. Each square in the figure corresponds to a voxel. The points and lines in the figure are colored according to their respective mesh. Some points are used twice to connect voxels, and to illustrate this these points are shown with both colors.

than the voxel sizes used in MRI scans to encompass more streamline points per voxel. To obtain the voxel coordinate of a streamline point p, we first calculate its position relative to the dataset by subtracting the minimum bound B_{min}, which is defined as the smallest coordinates contained in the dataset: $p_{rel} = p - B_{min}$. This ensures that each voxel has non-negative coordinates. The voxel coordinates v are then obtained by dividing p_{rel} by the voxel size s and taking the floor of the resulting vector: $v = \lfloor p_{rel}/s \rfloor$.

2.2   Generating Voxlines

Intuitively, grouping points of a streamline based on which voxel the points fall in, can be seen as dividing the streamline points into several parts when a streamline passes through multiple voxels. This can be seen in Fig. 1 where each streamline is split based on the voxel coordinate of each streamline point. In the rest of the paper, we will refer to these parts as "voxlines". A voxline is defined as a sequence of consecutive points of a streamline that is bounded by a voxel. In this context, consecutive means that there are no missing points between the minimum and maximum points of a group of streamline points. For example, points with streamline indices $\{6, 7, 8, 9\}$ would be considered consecutive, while $\{3, 4, 12, 13\}$ would not.

To generate voxlines making up the streamlines we can split each streamline in voxlines based on the voxel each point falls within. However, using only the points in each voxline would result in gaps between voxels when rendering the voxlines. Consider a streamline that is separated into two voxlines. If we render these voxlines as two distinct sets of lines, there will be a gap between them because no line is drawn between the end of the first voxline and the beginning of the second voxline. So defining the voxlines by using only the points that fall within the voxel would cause lines that cross the voxel borders to not be rendered. To address this issue, when generating the voxlines we add an additional point for every voxline that does not contain the final streamline point. This additional vertex corresponds to the next point in the streamline that falls outside of the voxel. By doing this, we fill all the gaps between the voxlines. This is illustrated in Fig. 1 by certain voxlines having an extra point outside of their voxel boundary and are thus part of both voxel meshes.

Algorithm 1: Voxline algorithm. Generates voxlines grouped by voxel coordinates from the input streamlines.

Data: Input data consist of a sequence *streamlines* where each element consists of a sequence of points, bound minimum B_{min} of the whole dataset and voxel size s blue.

$voxelset = \emptyset$ //Empty hashmap with voxel coordinates as keys and sets of voxlines as values

forall the *streamlines* $S = < p_0, p_1, ... p_{n-1} >$ **do**

   $v_p = null, i = 0, voxline = <>$

   while $i < n$ **do**

      $v_i = \lfloor (p_i - B_{min})/s \rfloor$

      if $v_p \neq v_i$ **then**

         if $i \neq n - 1$ **then**

            $voxline \leftarrow voxline + p_i$

         if *voxelset* **contains key** v_p **then**

            $voxelset[v_p] \leftarrow voxelset[v_p] \cup \{voxline\}$

         else

            $voxelset[v_p] \leftarrow \{voxline\}$

         $v_p = v_i, voxline = \emptyset$

      $voxline \leftarrow voxline + p_i$

      $i = i + 1$

   if $voxline \neq \emptyset$ **then**

      if *voxelset* **contains key** v_p **then**

         $voxelset[v_p] \leftarrow voxelset[v_p] \cup \{voxline\}$

      else

         $voxelset[v_p] \leftarrow \{voxline\}$

Result: Set of voxlines grouped by voxel coordinates.

Algorithm 1 implements the voxelization and voxeline generation described in Sects. 2.1 and 2.2. It generates the set of voxlines and groups them based on voxel coordinates by iterating over the streamlines. Each streamline is split into voxlines whenever a point is found with a different voxel coordinate than the previous point. Voxlines that do not contain the final point have an additional point added to ensure proper connectivity between voxlines.

2.3   Mesh Generation

We generate a mesh per voxel containing all voxlines part of the voxel, which can be rendered using the OpenGL 'lines' primitive. A mesh consists of vertices, each with a position and indices that define the render order of the lines. The mesh vertices are simply defined by the points of each voxline that fall within the voxel. We define the indices by concatenating each pair of consecutive voxline points within a voxel. The order of these pairs in the indices determines the rendering order of the line segments within a voxel. For now, we define this order simply based on the dataset order. Later, we will describe a more sophisticated view-dependent order in Sect. 2.6. Note that this method focuses solely on vertex positions and indices to determine the render order of the lines. Additional vertex information can be stored for cosmetic purposes, such as line identifiers, relative point indices, line tangents, and render flags. However, this additional data is not relevant for the render method as we are only concerned with vertex positions and the render order for improved transparency.

2.4   Render Order Accuracy

Transparency issues become most apparent when distant line segments are rendered in the incorrect order, specifically when line segments closer to the view are drawn before line segments further away. To address this, we divided the dataset into voxels, with each voxel representing the line segments passing through its bounds. Additional line segments were included to connect different voxel meshes. Prior to rendering the voxel, we sort them from back to front based on the current view. This sorting provides guarantees on the render order between any two line segments when the distance between any two consecutive streamline points is smaller than the voxel size. Line segments that are fully contained within different voxel bounds will be rendered in the correct back-to-front order. This holds true for the majority of line segments. Line segments that cross voxel bounds may have inaccurate render order relative to their immediate neighboring voxels. However, they are guaranteed to be in the correct order relative to line segments in non-neighboring voxels. Since we have not specifically ordered the lines within a voxel, the render order of line segments within a voxel will be inaccurate. Therefore, voxelization ensures that any two line segments have an accurate render order unless they are within the same voxel or one voxel apart for line segments crossing a voxel boundary, effectively resolving the most significant render order issues.

2.5   Improved Transparency Within Voxels

The primary issue remaining with the render order, and consequently transparency, is the inaccurate rendering order within a voxel. We can take advantage of the fact that voxelization divides the dataset into smaller datasets with similar properties to the original streamline data. Since both the original streamline dataset and each voxel consist of sets of lines composed of consecutive points, we can apply existing rendering techniques that approximate transparency for 3D line sets on a voxel level. Furthermore, sorting within voxels becomes more feasible because each voxel contains a smaller subset of the points. For this method, we have developed a solution that stores different streamline orders per voxel to improve transparency.

2.6   View-Dependent Line Order per Axis

To achieve more accurate transparency within voxels, we can sort the line segments based on the average position of the two endpoints and the current viewing direction. However, performing this sorting process every time the view changes would be computationally expensive, even after voxelization. Instead, we precompute line render orders for a set of viewing directions. When rendering, we select the precomputed "closest" viewing direction for improved transparency. The closest viewing direction is determined by calculating the distance between each stored sorting direction and the current viewing direction using the formula $s \cdot \frac{v-c}{||v-c||}$, where s represents the sorting direction, v is the voxel position, and c is the camera position.

This method is similar to imposter rendering, a technique that precomputes textures based on specific camera angles and then renders billboards instead of meshes for improved performance. However, instead of precomputing textures, we precompute line orders for a set of view directions. For the set of view directions, we use the six vectors defined by positive and negative unit vectors along each axis (X, Y, and Z). These six vectors are particularly well-suited as they are faster to sort than others, requiring no squared distance calculations. Additionally, these axis viewing directions align with axial, sagittal, and coronal projections, commonly used in tractography tools.

To sort the line segments for an arbitrary direction d, we can compare $\frac{(p_0 + p_1)}{2} \cdot d$ for each line segment, where p_0 and p_1 represent the two points making up the line segment. Since we only use unit vectors for d, we can simplify the comparison to a single vector component, comparing the non-zero component of d. Additionally, we optimize the sorting by using the sum of the two line positions instead of the average position. For example, sorting line segments for the direction $(0, 1, 0)$ would involve comparing the Y component of $p_0 + p_1$ for each line segment in the voxel. To find the orders of the negative unit vectors, we simply reverse the orders found for the positive unit vectors.

2.7   Evaluation

To evaluate our proposed approach, we generated a set of whole-brain tractograms from a single participant sourced from the Human Connectome Project [10]. A tractogram consisting of one million streamlines was constructed using multi-shell multi-tissue constrained spherical deconvolution in MRtrix (Fig. 2).

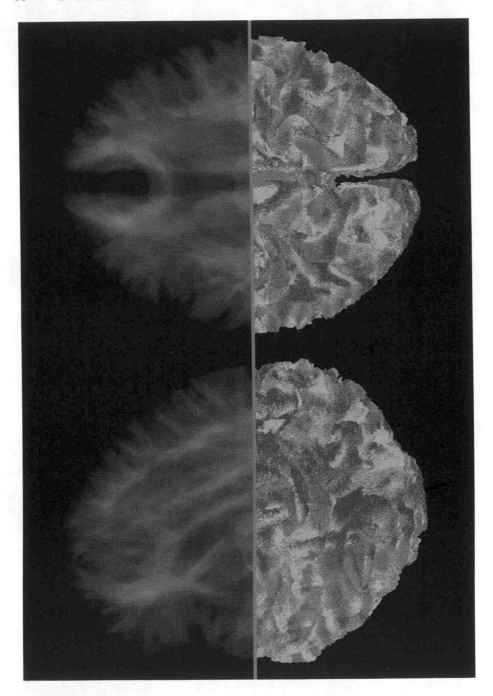

Fig. 2. This figure shows two different angles of a dataset consisting of one million streamlines rendered using our method with voxelization and axis view-dependent line orders. Each image is divided into two halves, with the left half rendered with 0.5% streamline opacity, and the right half rendered without transparency.

Subsequently, a smaller dataset was created using TractSeg [12], resulting in a tractogram with 140k streamlines. We implemented our render method in Neu-roTrace (gitlab.com/Besm/NeuroTrace), a tractography visualization tool developed by the authors. We compared the visual and performance results of our method with two popular tractography visualization tools that support transparency, namely MRtrix and TrackVis. We compare them to our method with only voxelization (basic) and the extension with axis based view-dependent line orders (axis). All results use voxel size 10mm^3, which through experimentation was found to give a balance between performance and visual quality for both individual bundles and whole brain tractograms.

3   Results

3.1   Qualitative

Figure 3 displays different transparency values for each method using a single dataset. MRTrix handles transparency by blending two renders of the dataset, one with a depth buffer and another without, based on the transparency value of the dataset [4]. While this approach preserves line render order, it makes it challenging to visualize the internal structures of the dataset. TrackVis renders transparent streamlines based on the given dataset, causing noticeable render order issues, particularly when streamline transparency is low.

In our method, by utilizing only voxelization (basic), we can observe more information about the streamlines deeper in the brain with minimal sorting inaccuracy. This is evident in the figure, for example, green superior longitudinal fasciculus fibers can be seen behind the red, short superficial fibers. Increasing transparency increases the visibility of the superior longitudinal fasciculus fibers without distorting the fact that the short superficial fibers are in front, while MRTrix and TrackVis have more difficulty conveying this information.

When applying axis sorting, the render order issues are slightly reduced. Although it may be challenging to discern in the image, axis method gives a more accurate representation of the render order compared to basic method. For instance comparing highlighted regions in Fig. 3 shows axis sorting bringing out certain red superficial fibers in front more clearly compared to the method without axis sorting.

3.2   Quantitative

Table 1 presents the performance comparison of our method with MRTrix and TrackVis on two tractography datasets: one with 140k streamlines and another with 1 million streamlines. It is worth noting that some tractography rendering methods include a preprocessing step that converts the dataset to a different format, aiming to reduce loading time for subsequent dataset loading. In our method, we intentionally omitted this preprocessing step to ensure a fair time comparison with MRTrix and TrackVis. Our implementation load the same TCK file format as MRTrix.

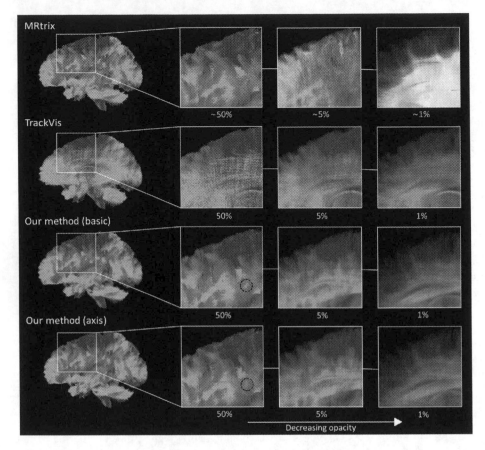

Fig. 3. Figure 3 shows a comparison of transparency between MRTrix, TrackVis, and our method, with just voxelization (basic) and with view-dependent internal voxel order (axis). For each method, we display the entire dataset with 50% transparency value from the sagittal plane and highlight a portion of the data with three opacity values: 50%, 5%, and 1%. Since MRTrix does not display opacity values, three comparable values were selected for the comparison.

4   Discussion

Voxelization and view-dependent line orders have shown promising results. The main benefits are from the voxelization step. View-dependent line order per voxel improve visual results at a computational cost, allowing for increased quality or larger voxel sizes with equivalent visual quality. Although we have not achieved the same performance as MRtrix, both our rendering and loading times are usable and faster than TrackVis. We believe that the decreased performance is caused by our implementation and is unrelated to the method outlined in this paper. In our implementation loading a dataset without voxelization has similar performance to loading with voxelization, indicating suboptimal performance

Table 1. Performance comparison between different methods with transparency. Loading time is measured from selecting the dataset (TCK/TRK file) to first render of the dataset. Performance is measured on a Windows Laptop with NVIDIA GeForce GTX 1650 and Intel(R) Core(TM) i7-9750H 2.60GHz CPU. TrackVis has additional loading time whenever modifying transparency value, which is shown in parenthesis.

Streamlines	Points	Method	Loading time (sec)	Render times (ms)
143,999	3,770,127	MRTrix	0.8	14
		TrackVis	19.8 (+2)	30
		our method (basic)	3.6	26
		our method (axis)	6.5	34
1,000,000	56,240,953	MRTrix	2.6	67
		TrackVis	283 (+26)	253
		our method (basic)	27	124
		our method (axis)	143	330

in dataset loading independent of our method. The tractography visualization tool we used to implement this method, NeuroTrace, is primarily focused on functionality and experimentation rather than performance optimization. We aim to improve performance in future work.

Currently, we manually determine the voxel size and whether to generate the view-dependent render orders. We are exploring the possibility of utilizing metadata of the dataset, such as step size and total point count, to automatically determine the parameters of our method. We are experimenting with varying transparency per streamline to highlight specific parts of the dataset and utilizing the benefits gained from voxelization in novel ways. We aim to explore these aspects in future work.

4.1 Different Sorting Orders

Regarding the set of precomputed view direction line orders, we have investigated two approaches: per-axis sorting and pseudo-random sorting. The per-axis method used in this work involves generating two orders per axis. For each voxel, we store line render orders in both ascending and descending order for each axis. The pseudo-random sorting approach involves sorting points based on pseudo-random directions that differ for each voxel. Per-axis sorting offers certain benefits mentioned previously, mainly performance, as axis directions do not require (squared) distance calculations while sorting on any other direction does. Additionally, these axis viewing directions align with axial, sagittal, and coronal projections, which are common viewing angles in tractography tools. However, we encountered an issue with voxelization and per-axis method where moving the camera tends to update certain lines of voxels simultaneously, causing visual jittering when rotating the viewing direction while using high streamline opacity (besm.gitlab.io/voxlines/videos). Using the pseudo-random direction mitigates this issue but has less accurate results for the most common viewing direction and is more computationally expensive.

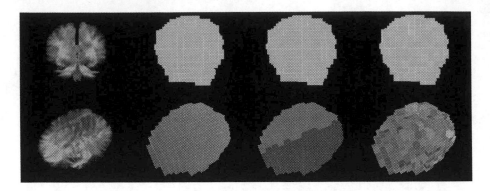

Fig. 4. This figure illustrates the voxel line ordering accuracy for two different camera angles. Each voxel is colored according to the absolute value of the direction for the RGB values. This direction differs per column. The second column displays the 'optimal sorting direction,' which is defined as the normalized direction from the camera to the voxel center. The third column shows the nearest voxel order direction using axis orderings. The fourth column shows the nearest voxel sorting direction using 64 pseudo-random sorting directions.

In Fig. 4, we demonstrate the difference between axis-based orders and pseudo-random direction orders. We can observe that the axis-based order is more accurate, when the dataset is viewed from one of the sides. Pseudo-random sorting performs worse in these cases. However, when considering arbitrary viewing directions, it outperforms the axis-based approach. Furthermore, when moving the camera, the pseudo-random method exhibits less visual jittering. However, the loading time for the pseudo-random method is worse. In future work, we aim to explore different sorting orders to find a balance between visual quality and performance.

Conclusion

In this work, we have proposed and implemented a novel transparency method that improves tractography visualizations, allowing the capturing of deeper structures in tractograms. We demonstrate improved transparency and comparable performance to existing tractography tools. Our method achieves this by providing a novel approximate transparency, which enhances the visibility of structures in the deeper regions of the brain compared to existing methods.

References

1. Everitt, C.: Nvidia corporation: order-independent transparency (2001). https://developer.download.nvidia.com/assets/gamedev/docs/OrderIndependent Transparency.pdf
2. Kanzler, M., Rautenhaus, M., Westermann, R.: A voxel-based rendering pipeline for large 3d line sets. IEEE Trans. Visual. Comput. Graph. **25**(07), 2378–2391 (01 2018). https://doi.org/10.1109/TVCG.2018.2834372
3. Kern, M., Neuhauser, C., Maack, T., Han, M., Usher, W., Westermann, R.: A comparison of rendering techniques for 3D line sets with transparency. IEEE Trans. Visual Comput. Graph. **27**(8), 3361–3376 (2021). https://doi.org/10.1109/TVCG. 2020.2975795
4. MRtrix3: Add support for visualizing tractography data (Issue #177). https://github.com/MRtrix3/mrtrix3/issues/177
5. Rheault, F., Houde, J.C., Descoteaux, M.: Visualization, interaction and tractometry: dealing with millions of streamlines from diffusion MRI tractography. Front. Neuroinform. **11** (06 2017). https://doi.org/10.3389/fninf.2017.00042
6. Salvi, M., Vaidyanathan, K.: Multi-layer alpha blending, pp. 151–158 (03 2014). https://doi.org/10.1145/2556700.2556705
7. Schultz, T., Sauber, N., Anwander, A., Theisel, H., Seidel, H.P.: Virtual klingler dissection: putting fibers into context. Comput. Graph. Forum **27**(3) (2008). https://doi.org/10.1111/j.1467-8659.2008.01243.x
8. Tax, C., et al.: Seeing more by showing less: orientation-dependent transparency rendering for fiber tractography visualization. PloS one **10**, e0139434 (10 2015). https://doi.org/10.1371/journal.pone.0139434
9. Tournier, J.D., et al.: Mrtrix3: A fast, flexible and open software framework for medical image processing and visualisation. NeuroImage **202**, 116137 (2019). https://doi.org/10.1016/j.neuroimage.2019.116137. https://www.sciencedirect.com/science/article/pii/S1053811919307281
10. Van Essen, D.C., Smith, S.M., Barch, D.M., Behrens, T.E., Yacoub, E., Ugurbil, K.: The wu-minn human connectome project: an overview. NeuroImage **80**, 62–79 (2013). https://doi.org/10.1016/j.neuroimage.2013.05.041. https://www.sciencedirect.com/science/article/pii/S1053811913005351, mapping the Connectome
11. Wang, R., Benner, T., Sorensen, A., Wedeen, V.: Diffusion toolkit: a software package for diffusion imaging data processing and tractography. In: Proceedings of the International Soc Mag Reson Med vol. 15 (01 2007)
12. Wasserthal, J., Neher, P., Maier-Hein, K.H.: Tractseg - fast and accurate white matter tract segmentation. NeuroImage **183**, 239–253 (2018). https://doi.org/10.1016/j.neuroimage.2018.07.070, https://www.sciencedirect.com/science/article/pii/S1053811918306864

Subnet Communicability: Diffusive Communication Across the Brain Through a Backbone Subnetwork

S. Shailja[1]([✉]), Jonathan Parlett[1], Abhishek Jeyapratap[1], Ali Shokoufandeh[1], Birkan Tunc[2], and Yusuf Osmanlioglu[1]

[1] Department of Computer Science, Drexel University, Philadelphia, PA, USA
`shailja@ucsb.edu`
[2] Children's Hospital of Philadelphia, Philadelphia, PA, USA

Abstract. One of the fundamental challenges in modern neuroscience is understanding the interplay between the brain's functional activity and its underlying structural pathways. To address this question, we propose a novel communication pattern called *subnet communicability*, which models diffusive communication between pairs of regions through a small, intermediary subnetwork of brain regions as opposed to spreading messages through the entire network. We demonstrate that subnet communicability strengthens coupling between the structural and functional connectomes better than previous models, including communicability. Over two large datasets, we show that the optimal subnetwork is consistent across the population. Subnet communicability provides new insights into structure-function coupling in the brain and offers a balance between redundancy in message passing and economy of brain wiring.

Keywords: Structure-function coupling · Connectomics · Neuroimaging · Communicability · Subnet Communicability

1 Introduction

The human brain is a complex network of interconnected neural elements that can be considered as an information processing network. At the macroscale, the human connectome maps the connectivity between brain areas [6]. Using magnetic resonance imaging (MRI), the connectivity between brain regions can be described in terms of structural relationships between gray matter regions that denote anatomical connectivity through white matter pathways [23], or functional relationships capturing statistical patterns of co-activation over time that correspond to communication between these regions [3]. Functional connectivity and rich network dynamics are influenced and constrained by anatomical connections and brain network topology [4,12]. Understanding the dynamics of functional interactions between brain regions with no direct anatomical connections [13] is an open challenge in modern neuroscience.

Several network communication models have been devised to explain the relationship between observed functional connectivity (FC) and underlying structural connectivity (SC) in the brain. By modeling the possible neural communication pathways shaped by structural connections, these models aim to describe the

M. Karaman et al. (Eds.): CDMRI 2023, LNCS 14328, pp. 104–117, 2023.
https://doi.org/10.1007/978-3-031-47292-3_10

polysynaptic interactions between anatomically unconnected brain regions [1]. Using a particular candidate model, each individual's structural connectivity matrix can be augmented into a communication matrix, i.e., simulated functional connectivity, that represents the connectivity between all pairs of regions in the brain [21]. This simulated functional connectome is then compared with the empirical functional connectome to quantify similarity.

One of the earliest models proposed was shortest-path, where indirect communication between region pairs occurs through a minimum number of intermediate hub regions [12]. When considered on a weighted connectome, a variation of shortest-path was suggested where messages travel through the path with strongest connectivity [24]. These were considered de facto communication models in the brain as the idea of communication through a single optimal path is in alignment with the established wiring economy of the brain [2]. However, the efficacy of shortest path was questioned since its calculation requires having a complete knowledge of the network topology, which can be considered implausible for local neural elements in the brain [22]. Additionally, since the model assumes communication happens through a single pathway, it lacks the redundancy that is necessary for robust communication.

To overcome these limitations, communication patterns such as path transitivity and search information were proposed. These models advocate communication through parallel pathways that detour around the shortest path [11] and demonstrate better structure-function coupling than the shortest-path model. A decentralized communication pattern called communicability [7,9] models communication occurring diffusively through all possible pathways in parallel. Communicability was recently shown to increase coupling between the structural and functional connectome better than other communication patterns including shortest-path, search information, and path transitivity [17,20,28]. Despite its success in better explaining structure-function coupling and accounting for redundancy, communicability contradicts the established economy of brain wiring as it does not restrict volume of information sent when flooding the entire network for communication.

In this study, we propose a novel network communication model called *subnet communicability*. Our proposed model aims to limit the redundancy of communicability and provide an efficient wiring economy while still offering a decentralized communication scheme. We achieve this by restricting diffusive communication to occur through a backbone subnetwork of a considerably smaller size, which is connected to the rest of the network. We systematically evaluate subnetworks of varying sizes and investigate the set of regions that constitute subnetworks with the highest structure-function coupling across individuals. We also analyze the contribution of functional systems to these subnetworks. Our proposed model offers insight into the underlying mechanisms used to process information in the brain and its biological neural signaling patterns.

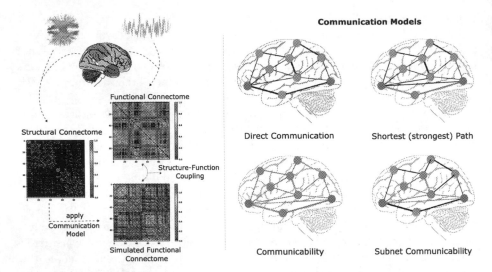

Fig. 1. Overview of the method: (left) Using the structural connectome of an individual, one of the communication models is applied to obtain a simulated functional connectome, which is then correlated with the positive functional connectome to calculate structure-function coupling. **(right)** Visual representation of explored communication models. Red nodes denote source and destination regions in the brain network. Blue edges represent the pathway that the message travels through, and edge thickness accounts for the strength of the structural connection. (Color figure online)

2   Methods

2.1   Dataset and Preprocessing

We evaluated our methods on 200 unrelated, healthy young adults (96 males) in the age range [22,35] from the S1200 Young Adult Open Access dataset of the Human Connectome Project (HCP) [25]. To test the generalizability of our approach, we repeated our experiments on 261 healthy individuals (140 males) in the age range [22,86] from the 1000Brains dataset [5]. Structural and functional connectomes used in our analysis were provided open source in [8,15], which were derived from diffusion-weighted MRI (dMRI) and resting-state functional MRI (fMRI) data. Connectomes were generated using the Schaefer atlas with 100 regions [18], where structural connectivity was obtained through probabilistic tracking with 10M streamlines and functional connectivity was obtained by calculating Pearson's correlation over the BOLD signal. The reader is referred to [8] and [15] for more details of the data processing pipeline.

2.2   Overview of Structure-Function Coupling

Taking the structural connectivity of a subject as the basis, we calculated simulated functional connectivity by using a communication model. We then calculated Pearson's correlation between the resulting simulated connectome and positive empirical functional connectome to quantify structure-function coupling (SFC). Higher correlation indicates better coupling (Fig. 1, left).

2.3 Communication Models

In our structure-function coupling analysis, we propose subnet communicability and compare it to three other communication models (Fig. 1, right). Each communication model is applied on a subject's SC matrix W, where W_{ij} is the strength of structural connections between region pairs.

Communicability. Communicability, which is the basis of our proposed model, utilizes a broadcasting approach where signals are simultaneously propagated through all possible regions in the network [9]. Unweighted communicability between nodes i, j is computed by calculating the number of walks between them scaled relative to path length k as $\text{Comm}_{ij} = \sum_{k=0}^{\infty} \frac{1}{k!}[W^k]_{ij}$. To account for the influence of connection strength in weighted SC matrices, weighted communicability is computed over the normalized W', where $W'_{ij} = W_{ij}/(\sqrt{s_i}\sqrt{s_j})$, and s_i is the strength of node i.

Subnet Communicability. Extending the definition of weighted communicability, which propagates a signal across all possible fronts, we propose subnet communicability which propagates a signal only through a subset of regions that constitute a backbone network. Given a graph $G = (V, E)$ and a subset of nodes $H \subseteq V$ to constitute a subnetwork, we compute subnet communicability between two regions $i, j \in V$ by first forming the subgraph composed of the nodes $H \cup \{i, j\}$ and their associated edges in E with corresponding SC matrix W_H. We then calculate connectivity by $\text{SubnetComm}_{ij} = \sum_{k=0}^{\infty} \frac{1}{k!}[W_H^k]_{ij}$. Repeating this process for all node pairs yields the subnet communicability matrix.

Shortest Path. The shortest path model routes information deterministically using a centralized strategy [21]. Given a weighted W, connectivity between i and j is given by $P_{ij} = e_{iu} + \ldots + e_{vj}$, the sum of edge weights in the strongest path between i and j, it is computed using an all pairs shortest path algorithm given W' as input, where $W'_{ij} = 1/W_{ij}$

Direct Communication. Used as a baseline for comparing the efficacy of the other communication models, this model accounts for communication happening only between regions that are anatomically connected to each other.

3 Results

SFC Analysis on Subnetwork Size: We first explored the SFC for subnet communicability using varying network sizes ranging from 1 to 75 regions across

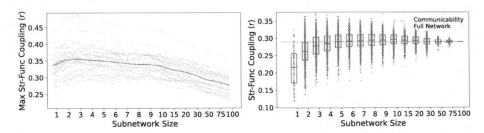

Fig. 2. SFC using subnet communicability for varying subnetwork sizes:
(left) Highest SFC achieved for each of 40 subjects across varying subnetwork sizes
are plotted in gray with their average plotted in purple, demonstrating a peak at subnetworks of size 3 and a decaying SFC with increasing subnetwork sizes. Subnetwork of
size 100 corresponds to the putative weighted communicability model **(right)** Distribution of SFC over randomly sampled subnetworks of varying sizes for a single subject
demonstrates higher SFC for certain subnetworks relative to standard communicability
utilizing the entire network (100 regions). (Color figure online)

a subset of the HCP dataset consisting of 40 individuals. We randomly sampled
nodes to constitute subnetworks 1000 times at each size (except for size 1 where
the number of possible subnetworks is 100), while ensuring that each node in the
graph has a connection to at least one of the nodes in the resulting subnetwork.
Evaluating the highest SFC scores for subjects across random samplings at each
size, we observed that SFC achieves a peak for subnetworks of size 3 with a
mean score of $r = 0.35$ across the subjects and steadily decays with increasing
subnetwork sizes, converging to $r = 0.27$ for standard communicability (Fig. 2,
left). For a single subject, SFC scores of all randomly sampled subnetworks
of varying sizes are also shown (Fig. 2, right). Initial results demonstrate that
higher SFC can be achieved using subnet communicability relative to standard
weighted communicability that uses the entire network.

Composition of Subnetworks: Having empirically determined the subnetwork size with peak SFC, we then analyzed the composition of subnetworks at
both the region and system levels. We first investigated brain regions that constitute networks achieving highest SFC. For each of the 200 individuals in the
HCP dataset, we randomly sampled 10,000 subnetworks of size 3 while ensuring
connectivity of all regions. We calculated SFC of subnet communicability utilizing each sampled subnetwork as a backbone and evaluated the consistency of
subnetwork regions based on how many times a region is included in the subnetwork achieving the highest SFC. As illustrated in Fig. 3, medial prefrontal cortex
(MPC) and orbital frontal cortex (OFC) regions were bilaterally over-represented
relative to the rest of the regions by several orders of magnitude (Left MPC =
89, right MPC = 98, left OFC = 68, right OFC = 72 occurrences). If region
occurrences in a subnetwork were based on random chance, only 6 occurrences
would be expected per region across all individuals, thus denoting the signifi-

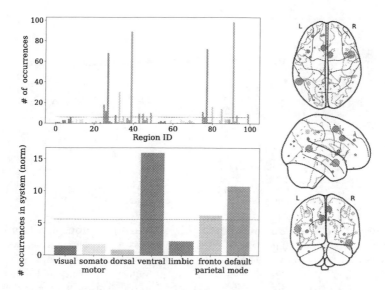

Fig. 3. Composition of subnetworks: (top left) Frequency of regions in the highest SFC subnetwork of size 3 across 200 subjects, where x-axis denote region IDs. Medial prefrontal cortex and orbital frontal cortex regions bilaterally occurred significantly more than the rest of the regions. **(bottom left)** Frequency of functional systems represented in the highest SFC subnetwork, normalized by system sizes. Default mode and ventral systems are disproportionately overrepresented relative to other systems. (Dashed red lines at the top and bottom indicate the number of times a region is expected to appear in networks across people if the occurrences were by random chance.) **(right)** Occurrences of regions expressed in proportion with the node radius over a brain image. (Color figure online)

cance of the MPC and OFC brain regions. We also noted that only 19 regions were above that threshold.

We then evaluated the representation of the seven functional systems [27] in the highest SFC achieving subnetworks by grouping occurrences of regions into systems followed by a normalization using system size. We observed that the ventral system and default mode network were over-represented while the remaining systems were under-represented (Fig. 3, bottom left). To demonstrate generalizability of results for varying network sizes, we repeated the experiments for subnetworks of size 5 and observed the same pattern at both region and systems levels (Fig. S1 in Appendix 1).

Comparison of Communication Models: Having established the consistency of certain brain regions in the highest SFC achieving subnetworks, we then explored how subnet communicability performs relative to other communication models. For this subnet communicability analysis, we used the same subnetwork across individuals consisting of bilateral MPC and right OFC regions. We also calculated simulated functional connectomes using the other three

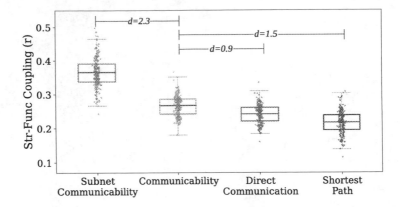

Fig. 4. Structure-function coupling of subjects using various communication patterns: Paired group differences were calculated between each communication pattern relative to the putative communicability model. Subnet communicability over a backbone network of 3 nodes achieves significantly higher SFC compared to communicability that uses the entire network for parallel communication. On the other hand, structural connectome without any model applied as well as shortest path achieved a lower SFC relative to communicability. All group differences were significant ($p < 10^{-6}$) after Bonferroni multiple comparison correction with large effect sizes (Cohen's d).

communication models. With the putative communicability model as a reference, statistical group comparison (paired t-test) revealed that subnet communicability ($\mu = 0.36$) achieved significantly higher SFC ($\mu = 0.27$) with a very large effect size (Cohen's $d = 2.3$, $p < 10^{-6}$) (Fig. 4). We further observed that direct communication ($\mu = 0.24$; $d = 0.9$, $p < 10^{-6}$) and shortest path ($\mu = 0.22$; $d = 1.5$, $p < 10^{-6}$) both achieve significantly lower SFC, confirming previous literature [17,20,28]. Once again, we observed the same pattern of group differences with subnetworks of size 5 (Fig.S2 in Appendix 1).

Centrality Analysis of Subnetworks: Finally, we investigated the network topology features of the highest achieving subnetworks by exploring node centrality measures. We calculated the betweenness (BC), subgraph (SC), eigenvector centralities (EC), and clustering coefficient (CC) of each region over the average SC across subjects. As 19 regions were observed to have representation higher than chance in the composition of subnetworks, we generated 4 sets of 19 regions having the highest network measure scores. We then compared the set of regions that appeared most frequently in subnetworks with these sets to quantify their overlap using the Dice coefficient. We observed low Dice coefficients for all measures (BC=0.32, SC=0.11, EC=0.05, CC=0.11), indicating that region representation in top subnetworks may not be related to the centrality metrics tested alone.

Replication of Results over a Second Dataset: We repeated the experiments over structural and functional data of 261 individuals from the 1000Brains dataset. The relationship between the size of the subnetwork and SFC demonstrated a trajectory similar to HCP data, where the highest SFC was achieved for networks of size 3–7, and it decreased by larger network sizes (Fig.S3 in Appendix 1). When comparing the nodes represented in the subnetwork of size 3 between HCP and 1000Brains datasets, although the individual regions that occurred most frequently were not the same, the nodes from visual, somatomotor, and dorsal systems were under-represented, and ventral and default mode networks were over-represented, which aligns with HCP results. However, we also observed that nodes from limbic and frontoparietal networks were over-represented in this second experiment (Fig.S4 in Appendix 1). Finally, repeating the experiment by using the top three most frequently occurring nodes on 1000Brains data, we observed the same pattern of subnet communicability achieving higher SFC than the other communication patterns with large effect sizes (Fig.S5 in Appendix 1). Overall, results were replicated to a large extent with some differences in the frequency of occurrences of individual regions.

4   Discussion

In this study, we proposed a novel communication model called subnet communicability that uses a considerably smaller subset of brain regions for diffusive message passing. We demonstrated that subnet communicability achieves better structure-function coupling relative to standard communicability over the entire network, which was previously shown to achieve the highest structure-function coupling through a diffusive message passing strategy [17,28].

The human brain's wiring is considered to minimize energy consumption [2] supporting communication models such as shortest path. These models deterministically route information using unique optimal paths, requiring global knowledge of network topology, which is deemed unlikely for local brain regions. Diffusive models like communicability, on the other hand, demonstrated higher structure-function coupling while also retaining redundancy in message passing. This approach, thus, ensures robustness in communication while requiring minimal information about network topology. However, communicability contradicts the well-established economy of communication in the brain due to its unrestricted message-passing scheme. Subnet communicability as we proposed, on the other hand, has demonstrated a balance between brain wiring economy and redundancy in message passing while requiring limited knowledge of network topology across neural elements by achieving highest SFC through a fairly small subnetwork of regions.

The over-representation of the default mode network (DMN) and ventral attention system in the subnetworks across subjects demonstrates a consistent pattern. DMN is known to be active during resting state [10] which might explain its central role in subnetworks indicated by the higher SFC relative to resting state function. The ventral attention system is known to orient attention towards internal state [26], which might further support its over-representation for SFC relative to resting state function where external stimuli are limited.

Finally, subnet communicability with the same subnetwork achieving higher SFC consistently across the entire cohort relative to other communication models highlights a biological basis for the choice of nodes that constitute this high-achieving subnetwork. The lack of a strong overlap between the frequency of nodes appearing in highest SFC achieving subnetworks and centrality scores of nodes, however, indicates that tested centrality measures are insufficient in explaining the underlying network topological mechanisms, requiring further investigation.

Although this study investigates the efficacy of subnet communicability over two high-quality datasets, certain limitations should be acknowledged. First, it is known that diffusion MRI has various inaccuracies in determining structural brain connectivity, especially at regions involving crossing or kissing fibers [14]. Second, experiments were carried out on a single atlas at a single parcellation resolution. Third, although the total number of samples investigated in the study is over 500, the datasets explored in the study only consisted of healthy samples, thus limiting the generalizability of results to the healthy. Also, the age range of samples did not include children and young adults, which limits the applicability of results to adults. Finally, prediction of functional interactions of brain regions from their structural connectivity is an inherently restricted problem due to differences of consistency of the data modalities [16].

5   Conclusions

Diffusive communication models were previously shown to explain SFC better in the brain and provide necessary redundancy in message passing despite contradicting the established economy of wiring in brain [17,19,28]. We have demonstrated that parallel communication when routed through a small subnetwork using subnet communicability explains SFC better than prior diffusive models while establishing a balance between redundancy and economy of brain. Subnet communicability presents an interesting communication pattern that warrants further exploration.

Acknowledgement. Data was provided by the Human Connectome Project, WU-Minn Consortium (Principal Investigators: David Van Essen and Kamil Ugurbil; 1U54MH091657) funded by the 16 NIH Institutes and Centers that support the NIH Blueprint for Neuroscience Research; and by the McDonnell Center for Systems Neuroscience at Washington University.

A   Appendix

A.1   Repeating Experiments Over a Subnetwork Consisting of Five Nodes on HCP Data

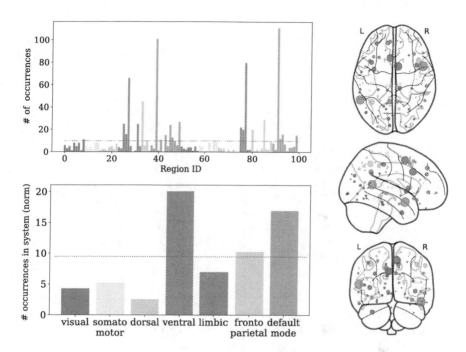

Fig. 5. Composition of subnetworks: (top left) Frequency of regions in the highest SFC subnetwork of size 5 across 200 subjects, where x-axis denote region IDs. Bilateral medial prefrontal cortices and orbital frontal cortices, and left lateral ventral prefrontal cortex regions occurred significantly more than the rest of the regions. **(bottom left)** Frequency of functional systems represented in the highest SFC subnetwork, normalized by system sizes. Default mode and ventral systems are disproportionately over represented relative to other systems. (Dashed red lines at top and bottom indicates the number of times a region is expected to appear in networks across people if the occurrences were by random chance.) **(right)** Occurrences of regions expressed in proportion with the node radius over a brain image. (Color figure online)

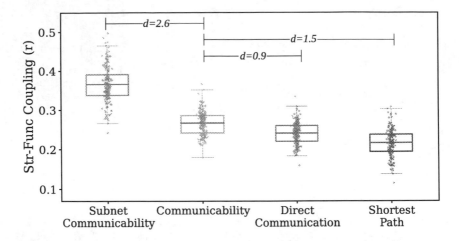

Fig. 6. Structure-function coupling of subjects using various communication patterns: Paired group differences were calculated between each communication pattern relative to the putative communicability model. Subnet communicability over a backbone network of 5 nodes achieves significantly higher SFC compared to communicability that uses the entire network for parallel communication. On the other hand, structural connectome without any model applied as well as shortest path achieved a lower SFC relative to communicability. All group differences were significant ($p < 10^{-6}$) after Bonferroni multiple comparison correction with large effect sizes (Cohen's d).

A.2   Replication Study: Repeating Experiments Over a Subnetwork Consisting of Three Nodes on 1000Brains Dataset

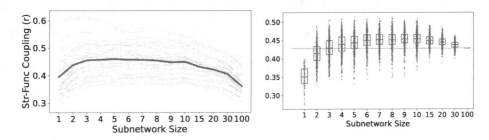

Fig. 7. SFC using subnet communicability for varying subnetwork sizes over 1000Brains dataset: (left) Highest SFC achieved for each of 40 subjects across varying subnetwork sizes are plotted in gray with their average plotted in purple, demonstrating a peak at subnetworks of size 3-7 and a decaying SFC with increasing subnetwork sizes. Subnetwork of size 100 corresponds to the putative weighted communicability model **(right)** Distribution of SFC over randomly sampled subnetworks of varying sizes for a single subject demonstrates higher SFC for certain subnetworks relative to standard communicability utilizing the entire network (100 regions).

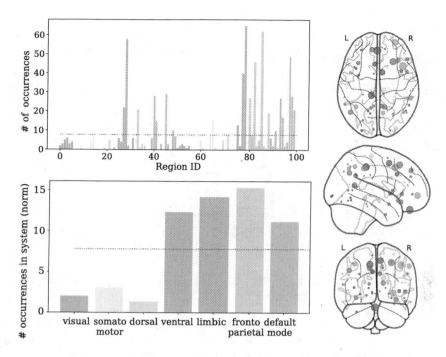

Fig. 8. Composition of subnetworks over 1000Brains dataset: (top left) Frequency of regions in the highest SFC subnetwork of size 3 across 261 subjects, where x-axis denote region IDs. Salient Ventral Attention Med-2 region on left hemisphere and Limbic Orbitofrontal Cortext-1 and PFCl-4 regions on the right hemispheres are top three most frequently chosen regions to constitute the subnetwork. **(bottom left)** Frequency of functional systems represented in the highest SFC subnetwork, normalized by system sizes. Limbic, frontoparietal, default mode, and ventral systems are disproportionately over represented relative to other systems. (Dashed red lines at top and bottom indicates the number of times a region is expected to appear in networks across people if the occurrences were by random chance.) **(right)** Occurrences of regions expressed in proportion with the node radius over a brain image.(Color figure online)

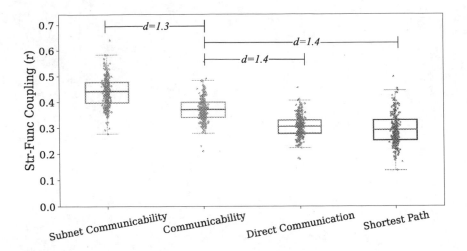

Fig. 9. Structure-function coupling of subjects using various communication patterns on 1000Brains dataset: Paired group differences were calculated between each communication pattern relative to the putative communicability model. Subnet communicability over a backbone network of three nodes achieves significantly higher SFC compared to communicability that uses the entire network for parallel communication. On the other hand, structural connectome without any model applied as well as shortest path achieved a lower SFC relative to communicability. All group differences were significant ($p < 10^{-6}$) after Bonferroni multiple comparison correction with large effect sizes (Cohen's d).

References

1. Avena-Koenigsberger, A., Misic, B., Sporns, O.: Communication dynamics in complex brain networks. Nat. Rev. Neurosci. **19**(1), 17–33 (2018)
2. Bullmore, E., Sporns, O.: The economy of brain network organization. Nat. Rev. Neurosci. **13**(5), 336–349 (2012)
3. Cabeza, R., Kingstone, A.: Handbook of Functional Neuroimaging of Cognition. MIT Press, Cambridge (2006)
4. Calamante, F., Smith, R.E., Liang, X., Zalesky, A., Connelly, A.: Track-weighted dynamic functional connectivity (TW-DFC): a new method to study time-resolved functional connectivity. Brain Struct. Funct. **222**(8), 3761–3774 (2017)
5. Caspers, S., et al.: Studying variability in human brain aging in a population-based German cohort-rationale and design of 1000brains. Front. Aging Neurosci. **6**, 149 (2014)
6. Craddock, R.C., et al.: Imaging human connectomes at the macroscale. Nat. Methods **10**(6), 524–539 (2013)
7. Crofts, J.J., Higham, D.J.: A weighted communicability measure applied to complex brain networks. J. R. Soc. Interface **6**(33), 411–414 (2009)
8. Domhof, J.W.M., Jung, K., Eickhoff, S.B., Popovych, O.V.: Parcellation-based structural and resting-state functional brain connectomes of a healthy cohort (v1.1) (2022). https://doi.org/10.25493/NVS8-XS5, https://search.kg.ebrains.eu/instances/f16e449d-86e1-408b-9487-aa9d72e39901

9. Estrada, E., Hatano, N.: Communicability in complex networks. Phys. Rev. E **77**(3), 036111 (2008)

10. Fox, M.D., Snyder, A.Z., Vincent, J.L., Corbetta, M., Van Essen, D.C., Raichle, M.E.: The human brain is intrinsically organized into dynamic, anticorrelated functional networks. Proc. Natl. Acad. Sci. **102**(27), 9673–9678 (2005)

11. Goñi, J., et al.: Resting-brain functional connectivity predicted by analytic measures of network communication. Proc. Natl. Acad. Sci. **111**(2), 833–838 (2014)

12. Honey, C.J., et al.: Predicting human resting-state functional connectivity from structural connectivity. Proc. Natl. Acad. Sci. **106**(6), 2035–2040 (2009)

13. Honey, C.J., Thivierge, J.P., Sporns, O.: Can structure predict function in the human brain? Neuroimage **52**(3), 766–776 (2010)

14. Jones, D.K.: Challenges and limitations of quantifying brain connectivity in vivo with diffusion MRI. Imaging Med. **2**(3), 341 (2010)

15. Jung, K., Eickhoff, S.B., Popovych, O.V.: Parcellation-based structural and resting-state functional whole-brain connectomes of 1000brains cohort (v1.1) (2022). https://doi.org/10.25493/8XY5-BH7, https://search.kg.ebrains.eu/instances/3f179784-194d-4795-9d8d-301b524ca00a

16. Osmanlıoğlu, Y., Alappatt, J.A., Parker, D., Verma, R.: Connectomic consistency: a systematic stability analysis of structural and functional connectivity. J. Neural Eng. **17**(4), 045004 (2020)

17. Osmanlıoğlu, Y., et al.: System-level matching of structural and functional connectomes in the human brain. Neuroimage **199**, 93–107 (2019)

18. Schaefer, A., et al.: Local-global parcellation of the human cerebral cortex from intrinsic functional connectivity MRI. Cereb. Cortex **28**(9), 3095–3114 (2018)

19. Seguin, C., Jedynak, M., David, O., Mansour L, S., Sporns, O., Zalesky, A.: Communication dynamics in the human connectome shape the cortex-wide propagation of direct electrical stimulation. bioRxiv, pp. 2022–07 (2022)

20. Seguin, C., Sporns, O., Zalesky, A., Calamante, F., et al.: Network communication models narrow the gap between the modular organization of structural and functional brain networks. Neuroimage **257**, 119323 (2022)

21. Seguin, C., Tian, Y., Zalesky, A.: Network communication models improve the behavioral and functional predictive utility of the human structural connectome. Netw. Neurosci. **4**(4), 980–1006 (2020)

22. Seguin, C., Van Den Heuvel, M.P., Zalesky, A.: Navigation of brain networks. Proc. Natl. Acad. Sci. **115**(24), 6297–6302 (2018)

23. Sporns, O., Tononi, G., Kötter, R.: The human connectome: a structural description of the human brain. PLoS Comput. Biol. **1**(4), e42 (2005)

24. Van Den Heuvel, M.P., Kahn, R.S., Goñi, J., Sporns, O.: High-cost, high-capacity backbone for global brain communication. Proc. Natl. Acad. Sci. **109**(28), 11372–11377 (2012)

25. Van Essen, D.C., et al.: The human connectome project: a data acquisition perspective. Neuroimage **62**(4), 2222–2231 (2012)

26. Vossel, S., Geng, J.J., Fink, G.R.: Dorsal and ventral attention systems: distinct neural circuits but collaborative roles. Neuroscientist **20**(2), 150–159 (2014)

27. Yeo, B.T., et al.: The organization of the human cerebral cortex estimated by intrinsic functional connectivity. J. Neurophysiol. (2011)

28. Zamani Esfahlani, F., Faskowitz, J., Slack, J., Mišić, B., Betzel, R.F.: Local structure-function relationships in human brain networks across the lifespan. Nat. Commun. **13**(1), 1–16 (2022)

Fast Acquisition for Diffusion Tensor Tractography

Omri Leshem[1]($\boxtimes$), Nahum Kiryati[2], Michael Green[3], Ilya Nelkenbaum[3], Dani Roizen[3], and Arnaldo Mayer[3]

[1] School of Electrical Engineering, Tel-Aviv University, Tel-Aviv, Israel
omrileshem@mail.tau.ac.il
[2] The Manuel and Raquel Klachky Chair of Image Processing, School of Electrical Engineering, Tel-Aviv University, Tel-Aviv, Israel
[3] Diagnostic Imaging, Sheba Medical Center, Affiliated to the School of Medicine, Tel Aviv University, Tel-Aviv, Israel

Abstract. Diffusion tensor tractography is a powerful method for in-vivo white matter mapping. Its implementation involves long scanning sessions to capture local diffusion orientations, followed by tedious post-processing to generate accurate tracts. While some initial research was conducted to reduce the number of required gradient directions, the current state-of-the-art still considers acquisition protocol acceleration and automatic tract segmentation as two separate tasks. We aim at optimizing the whole workflow, from acquisition to tract segmentation. We propose a collaborative neural framework for diffusion-encoding color map denoising and white matter tract segmentation. It generates high-quality white matter tracts using DWI acquired for a small number of diffusion-encoding gradient directions (GDs), thus minimizing acquisition and post-processing time. The proposed method is first validated on the high-angular resolution (270 GDs) HCP dataset using a novel spherical k-means method to select a subset of 16 quasi-uniformly distributed GDs. Further validation is provided for a prospective clinical dataset of 10 cases acquired at both 16 and 64 GDs. Encouraging experimental results are obtained using several state-of-the-art neural architectures and training loss functions.

Keywords: Diffusion Tensor MRI · Tractography · Denoising · Segmentation · Deep Learning

1 Introduction

Diffusion tensor (DT) tractography is a powerful tool for in-vivo brain white matter (WM) mapping. Nevertheless, the time required to acquire diffusion-weighted images (DWI), and the subsequent tedious generation of accurate white matter (WM) tracts, remain two major pain points. Acquisition time is proportional to the number of diffusion-encoding gradient directions (NDGD) implemented

© The Author(s), under exclusive license to Springer Nature Switzerland AG 2023
M. Karaman et al. (Eds.): CDMRI 2023, LNCS 14328, pp. 118–128, 2023.
https://doi.org/10.1007/978-3-031-47292-3_11

in the scanning protocol [12]. At least six non-co-linear gradient-weighted scans ($B \neq 0$), together with a reference one ($B = 0$), are required to fit the tensor model [2]. However, increasing the NDGD or the number of excitations (NEX) improves the SNR and robustness of the computed DT [12]. Higher NDGD also leads to superior tractography as diffusion is sampled at a higher angular resolution [20]. In [10], a convolutional neural network (CNN) is trained to transform DWI images directly into DTI maps, hence bypassing tensor fitting. High-quality DTI maps and deterministic tractography are claimed for an NDGD as low as 6, but validation is limited to synthesized low-NDGD data. Moreover, tractography is quantitatively assessed by comparing the reconstructed fibers' number with the ground truth, rather than spatial overlap using e.g. the Dice index.

In [16], CNNs are used to denoise six DWI volumes sampled along selected diffusion-encoding directions, while the ground truth (GT) is synthesized from tensors fitted to additional data. A 3.5-4 fold acceleration is claimed. In [17], Self-supervised CNN-based denoising is performed, eliminating the need for additional high-SNR training data. Specifically, several DWI repetitions are synthesized (using the DT model) with identical image domain contrasts but independent noise samples. Again, tract-level validation for the proposed method is not provided. In [19] a lower NDGD dataset is synthesized from high angular resolution HCP data (NDGD = 270) by interpolation of spherical harmonic functions. The purpose was to simulate a clinical dataset with a "standard" number of diffusion gradients (NDGD = 32), not to accelerate acquisition by minimizing the NDGD.

Generating accurate subject-specific tractography from the DTI maps still requires substantial work to seed, draw, and manually remove tract outliers. In recent years, CNN architectures have been very successful also at segmenting WM tracts from DWI or DTI maps possibly complemented by anatomical scans. The TractSeg network is proposed in [19] to directly segment tracts from fiber orientation distribution function (fODF) peaks. Tractseg was validated on the Human Connectome Project dataset. In [1] a multi-sequence CNN workflow was proposed for joint 3-D registration and segmentation of WM tracts. The method, applied to pairs of T1-weighted (T1w) and directionally encoded color (DEC) maps, was validated on HCP as well as a real pre-surgical dataset. We observe that tract segmentation prior art described above only considers high NDGD data, with at least 64 directions, while the DTI acceleration works did not consider automatic tract segmentation. In other words, the current state-of-the-art treats acquisition protocol acceleration and automatic tract segmentation as two separate problems.

Aiming at optimizing the whole workflow from acquisition to tract segmentation, this work proposes a neural framework for DEC map enhancement and automatic tract segmentation. It allows for the automatic segmentation of high-quality WM tracts generated from DWI volumes acquired at a small NDGD, thus significantly reducing acquisition and post-processing time. Specifically, the major contributions of this paper are:

- A fully automatic collaborative framework that, given input DEC maps computed from low-NDGD DWI volumes, generates high-quality segmentation of WM tracts.
- A novel spherical k-means method for the selection of approximately uniform diffusion-encoding gradient direction (DGD) subsets out of a high NDGD set for training of neural DEC map denoising.
- A novel diffusion loss for neural training of DEC maps denoising.

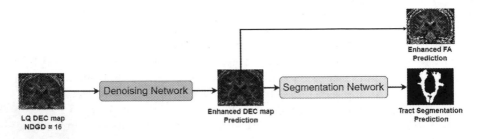

Fig. 1. The proposed framework. Collaborative training is presented in subsection 3.1.

2   Methods

2.1   Denoising and Segmentation Pipeline

Figure 1 shows the proposed neural pipeline for denoising and tract segmentation from low NDGD DEC maps. For DEC maps denoising we used the DnCNN3D architecture [11]. DnCNN3D is trained using a novel diffusion loss detailed below, with comparison to classical L_1 loss. For the subsequent tract segmentation, two architectures are considered: (1) a 4-layer Unet-3D [3] with a feature map size of 16 and group numbers set to 8. Unet-3D is trained using Dice loss [7]. (2) Swin-UNETR [15], based on a state-of-the-art shifted window visual image transformer encoder that extracts features at 4 different resolutions. The encoder is connected to a CNN-based decoder at each resolution by skip connections. Swin-UNETR is trained using the Dice-cross-entropy loss.

2.2   NDGD Reduction

Teaching the proposed framework to work with low NDGD can be done in two ways: One possibility, explored in prior art, is to generate the low-NDGD dataset from the full set of high NDGD DWI images containing high angular resolution information. The second is to acquire real pairs of scans at low and high NDGD. While the real data approach can learn a more realistic mapping between low and high NDGD DTI maps, it is less practical to deploy as more real acquisitions are needed. However, interpolation methods or DWI synthesis from the DT model may result in smoothed low-NDGD DTI maps due to interpolations.

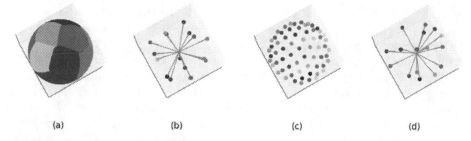

Fig. 2. (a) clustering of 10^5 GD on the unit sphere for $k = 16$. Each color represents the GDs assigned to the same cluster; (b) The 16 GDs selected by spherical k-means of the initial set from (a); (c) an initial set of 64 GDs; (d) following reduction from 64 to 16 GDs selected by spherical k-means.

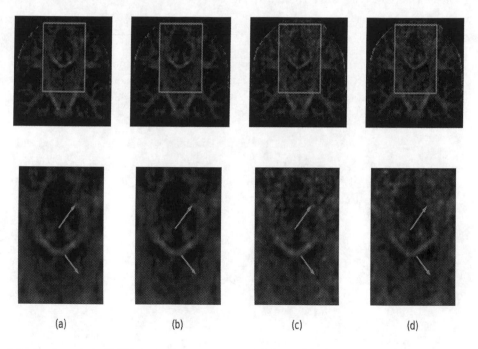

Fig. 3. A sample DEC map slice reconstructed from (a) data acquired at NDGD = 64; (b) k-means selected NDGD = 16 out of 64 GD; (c) NDGD = 16 generated by interpolation from 64 GD as per [19]; (d) data acquired at NDGD = 16. We note that both (b) and (c) are associated with the same GDs.

Instead of interpolating the whole NDGD set, utilizing complete angular information, we propose to select a subset consisting of specific gradient directions (GDs) that offers pseudo-optimal angular coverage. For this purpose, we perform spherical k-means clustering [5] of the initial GD set on the unit sphere, where k is the desired low NDGD. For each cluster, the GD closest to the centroid, in terms of cosine distance, is selected. Using the cosine distance in the spherical k-means algorithm is natural since, unlike the L1 or L2 norms, the cosine distance effectively measures angular dissimilarity [5]. For illustration, the cluster division of 10^5 GDs on the unit sphere is shown in Fig. 2(a) for $k = 16$. Each color on the sphere represents the GD associated with the same cluster out of which the 16 selected GDs are shown in (b). In (c), an initial set of 64 GDs is shown before its reduction to 16 GDs, selected by spherical k-means (d). A sample DEC map slice reconstructed from (a) data acquired at $NDGD = 64$, (b) data acquired at $NDGD = 16$, (c) k-means selected $NDGD = 16$ out of 64 GD, and (d) $NDGD = 16$ generated by interpolation from 64 GD as per [19], is shown in Fig. 3. We observe that the k-means selection (c) leads to higher visual similarity with the data acquired at $NDGD = 16$ (b), while the interpolation approach (d) suffers from apparent over-smoothing compared to (b), not representative of true low NDGD acquisition.

2.3   Denoising Loss Function

The DEC map contains all the information required to generate DTI tractography. Its voxels' RGB values are determined by the element-wise product of the fractional anisotropy (FA) by the (absolute value of the) elements of the principal direction of diffusion (PDD) vector map [13].

$$RGB = FA \cdot [PDD] \tag{1}$$

Here [PDD] denotes a vector whose elements are the absolute values of the PDD vector. The PDD vector is a unit vector hence [PDD] is also unit vector. Therefore,

$$FA = \|RGB\|_2 \tag{2}$$

$$[PDD] = RGB/FA \tag{3}$$

Given our goal to preserve the information of both FA and PDD in the denoised DEC image, we introduce a new loss function, $\mathcal{L}_{diff}$, called the diffusion loss:

$$\mathcal{L}_{diff} = w_1 \cdot \left(1 - [PDD]_{GT}^T [PDD]_{pred}\right) + \frac{w_2}{\sqrt{3}} \cdot \|FA_{GT} - FA_{pred}\|_1 \tag{4}$$

where

$$w_1 + w_2 = 1$$

Here the *pred* and *GT* subscripts stand for prediction and ground-truth, respectively. FA and PDD values are extracted from the RGB maps (predicted and ground-truth). For the FA values, we employ the average L_1 distance as the loss metric. As for PDD, which represents a unit vector field, we utilize the average cosine distance for angle comparison. Preserving angular similarity between the predicted and true PDDs is crucial, especially considering that the PDD plays a vital role in estimating the orientation of tracts. Therefore, the cosine distance is ideal for this purpose, as it directly measures the angular dissimilarity between PDDs. The loss components are combined by weight coefficients w_1 and w_2 are empirically set to 0.57 and 0.43, respectively. At voxels with degenerate FA (typically about 0.3% of the brain voxels), an arbitrary PDD of (1, 0, 0) is assigned as an arbitrary numerical workaround, resolving the division by FA = 0 in Eq. 3.

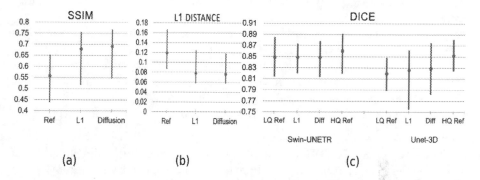

Fig. 4. Average value and range for: (a) the DEC images SSIM, (b) FA images L_1 distance, and (c) the segmented bilateral motor tracts Dice index in comparison to GT, for the HCP 20 test cases, with the Swin-UNETR and Unet-3D architectures. The Ref label stands for 16 GDs DEC map data generated by k-means selection compared to full 270 GDs DEC map. HQ and LQ references stands for the high/low NDGD HCP data, respectively. The L_1 and diffusion labels stand for the loss type used in training.

3   Experiments

Datasets. The Human Connectome Project (HCP) dataset [18] was used in the experiments involving a synthetic NDGD reduction. It consists of 105 cases (NDGD = 270) comprising 74 training, 11 validation, and 20 test cases. We followed the steps described in [19] to create ground truth segmentation for the bilateral motor tracts. NDGD reduction using the proposed spherical k-means approach was performed to select a subset of 16 GDs. An additional dataset comprising 10 real pairs of high and low NDGD DWI scans was also created to observe real acquisition-related effects of NDGD reduction. The data was acquired at Sheba Medical Center on a Philips 1.5T Ingenia scanner. DT, DEC and FA maps were generated by DIPY [6]. Contrack [14] was used to generate the bilateral motor tracts GT by probabilistic tractography. The dataset was aligned to MNI space (same template as in [19]) using FSL-Flirt [8,9]. A Binary brain extraction mask was computed and applied for each case using ANTsPyNet [4]. A k-fold scheme with $k = 5$ was implemented.

3.1   Training

DnCNN3D - Unet3D Scenario. A cascade training scheme with a joint loss was implemented for the denoising DnCNN3D and segmentation Unet-3D networks. The segmentation loss component consisted of Dice loss. For the denoising component, we compare two alternative loss functions: the proposed diffusion loss and the average L_1 loss. Both loss functions were applied with a binary brain mask to focus the evaluation on brain regions. To ensure comparability, we normalized the values of both the average L_1 loss and the diffusion loss to be in the interval $[0, 1]$. In the joint loss function, the denoising part was weighted equally for both the L_1 loss and diffusion loss, each contributing 0.6 to the overall loss. The segmentation part, measured by the Dice loss, was given a weight of 0.4. For the clinical dataset training, both DnCNN3D and UNet3d underwent transfer learning from the models previously trained in HCP experiments.

DnCNN - Swin-UNETR Scenario. Here the denoising and segmentation networks were not jointly trained due to the high memory requirements of Swin-UNETR. For DnCNN3D, the model trained in the previous scenario was used. The training data was forwarded through the parameter-frozen DnCNN3D to generate training inputs for Swin-UNETR which underwent self-supervised learning as in [15]. Although denoising and segmentation networks were trained separately, care was taken to keep the same fold partitions in both networks so that each of the networks was tested on the same never-seen-before cases.

3.2   Results

HCP Dataset. The plots in Fig. 4 give the average value and range for (a) the DEC images SSIM, (b) the FA images L_1 distance, and (c) the bilateral motor tracts Dice in comparison to GT, for the HCP 20 test cases. For DEC and FA, we observe a substantial improvement from pre to post-denoising SSIM (a) and L_1 distance (b), respectively. The same trend is observed for Unet3D segmentation Dice (c). Although Dice for Swin-UNETR segmentation outperforms UNET3D, it seems insensitive to prior denoising on the HCP dataset. The results also suggest that diffusion loss training slightly outperforms L_1 in the experiments.

Clinical Dataset. The quantitative results for the 5-fold validation on the 10 clinical cases dataset are shown in Fig. 5: (a) average SSIM on DEC map, (b) average L_1 distance on FA map, and (c) Dice for bilateral motor tracts. All indices are computed for two losses (diffusion and L_1). Dice is computed for two segmentation models, Unet3D and Swin-UNETR. For all indices, reference values on low-quality (LQ), not-denoised images, are given. For Dice, an additional high-quality reference result (HQ), computed on 64 GD, is also given.

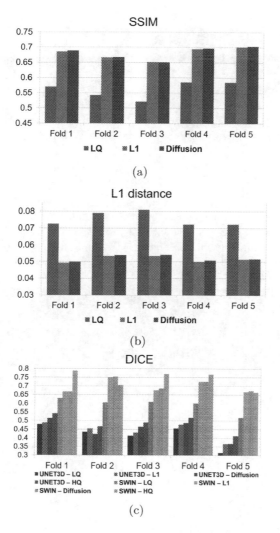

Fig. 5. Quantitative results: (a) average SSIM on DEC map; (b) average L_1 distance on FA map; (c) Dice (bilateral motor tracts). Indices are computed for diffusion and L_1 losses. Dice is computed for Unet3D and Swin-UNETR architectures. LQ are reference values on low-quality, not-denoised, images. For Dice, HQ are high-quality reference values for 64 GD.

Here again, a major improvement in the quantitative indices is observed for all folds between their values computed on the real low-NDGD images, acquired at NDGD = 16, before and after denoising. The average SSIM on DEC maps rises from 0.560 to 0.680 (L_1 loss) and 0.681 (diffusion loss). The average FA L_1 distance decreases from 0.075 to 0.051 (L_1 loss) and 0.052 (diffusion loss). The average Dice coefficient increases from 0.42 to 0.44 for the DnCNN3D-Unet-3D (denoising-segmentation) combination, and to 0.70 for the DnCNN3D-Swin-

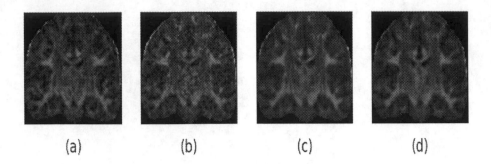

Fig. 6. Sample denoising results for a coronal DEC map reconstructed from: (a) NDGD = 64; (b) NDGD = 16; (c) NDGD = 16 denoised by DnCNN3D with diffusion loss; (d) NDGD = 16 denoised by DnCNN3D with L1 loss. It is evident that both predictions, (c) and (d), are significantly cleaner than the original NDGD = 16 input (b). Note that the predictions (c) and (d) are visually similar in this example.

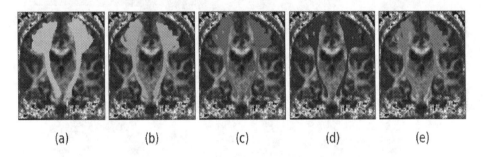

Fig. 7. Sample qualitative segmentation results for the bilateral motor tracts in overlay with FA for: (a) GT; (b) NDGD = 64; (c) NDGD = 16 denoised with diffusion loss; (d) NDGD = 16 with L_1 loss. Segmentation by Swin-UNETR in (b-d).

UNETR (denoising-segmentation) combination, approaching the average Dice of 0.74 obtained for Swin-UNETR segmentation on real 64 GD data (HQ).

We observe that although most of the Dice improvement is obtained for Swin-UNETR segmentation, the average Dice using the same segmentation but no prior denoising falls to 0.59. Such a clear sensitivity of the segmentation Dice to prior denoising was not observed for the HCP dataset with simulated NDGD reduction, suggesting the importance of acquiring real low-NDGD scans for validation completeness. This also suggests that some network pairs for consecutive denoising-segmentation may be more synergistic than others, with DnCNN3D-Swin-UNETR leading to the best results. Again, the diffusion loss slightly outperformed L_1 loss in almost every experiment.

Sample qualitative denoising results are shown in Fig. 6 for a coronal DEC map computed for: (a) NDGD = 64, (b) NDGD = 16, (c) NDGD = 16 denoised by DnCNN3D with diffusion loss, and (d) NDGD = 16 denoised by DnCNN3D with L_1 loss. Sample segmentation results are shown in Fig. 7 for the bilateral motor tracts overlaid with FA for: (a) the GT (probabilistic tractography) for NDGD = 64, (b) Swin-UNETR for NDGD = 64,(c) DnCNN3D-Swin-UNETR for NDGD = 16 and diffusion loss, (d) same as (c) but using L_1 loss instead, and (e) Swin-UNETR segmentation for NDGD = 16 without any denoising. We observe that, in agreement with the quantitative results, the DnCNN3D-Swin-UNETR combination for NDGD = 16 (c,d) substantially improves segmentation quality in comparison to Swin-UNETR segmentation for NDGD = 16 without denoising (e), approaching the Swin-UNETR segmentation for NDGD = 64(b).

4   Discussion and Conclusions

We presented a collaborative neural framework for fully automatic denoising and tract segmentation for fast DTI acquisitions. It was successfully validated on the Human Connectome Project (HCP) dataset and on a unique dual acquisition prospective 10 case clinical dataset. In the context of the (scientific) HCP dataset, we presented a novel method for reducing the NDGD (number of diffusion-encoding gradient directions). Unlike previous works, the reduction is by selection of a subset of GDs (gradient directions) using spherical k-means clustering. On the clinical dataset, the method provided major quantitative improvements of the DEC and FA maps. It demonstrated a promising ability to generate high-quality tracts from merely 16 GD data acquired on a 1.5T magnet at 4 min scan-time (reduced from 19). This may prove especially useful for intra-operative navigation, where the magnetic field and the scanning time are both limited. Moreover, even faster DTI acquisitions may be possible by applying the proposed methods to 3T diagnostic scanners, improving both patient experience and scanner throughput. In future research, extensive validation on larger clinical datasets will be performed, seeking the NDGD that best balances between scan time reduction and image quality, for different imaging scenarios e.g. diagnostic imaging and neurosurgical planning.

References

1. Barzilay, N., Nelkenbaum, I., Konen, E., Kiryati, N., Mayer, A.: Neural registration and segmentation of white matter tracts in multi-modal brain MRI. In: Karlinsky, L., Michaeli, T., Nishino, K. (eds.) Computer Vision – ECCV 2022 Workshops. ECCV 2022. LNCS, vol. 13803. Springer, Cham (2022). https://doi.org/10.1007/978-3-031-25066-8_12
2. Basser, P.J., Mattiello, J., LeBihan, D.: MR diffusion tensor spectroscopy and imaging. Biophys. J . **66**(1), 259–267 (1994)
3. Çiçek, Ö., Abdulkadir, A., Lienkamp, S.S., Brox, T., Ronneberger, O.: 3D U-Net: learning dense volumetric segmentation from sparse annotation. In: Ourselin, S., Joskowicz, L., Sabuncu, M.R., Unal, G., Wells, W. (eds.) MICCAI 2016. LNCS, vol. 9901, pp. 424–432. Springer, Cham (2016). https://doi.org/10.1007/978-3-319-46723-8_49

4. Cullen, N.C., Avants, B.B.: Convolutional neural networks for rapid and simultaneous brain extraction and tissue segmentation. In: Spalletta, G., Piras, F., Gili, T. (eds.) Brain Morphometry. Neuromethods, vol. 136, pp. 13–34. Springer, New York (2018). https://doi.org/10.1007/978-1-4939-7647-8_2

5. Dhillon, I.S., Modha, D.S.: Concept decompositions for large sparse text data using clustering. Mach. Learn. **42**, 143–175 (2001)

6. Garyfallidis, E., et al.: Dipy, a library for the analysis of diffusion MRI data. Front. Neuroinform. **8**, 8 (2014)

7. Huang, Q., Sun, J., Ding, H., Wang, X., Wang, G.: Robust liver vessel extraction using 3D U-net with variant dice loss function. Comput. Biol. Med. **101**, 153–162 (2018)

8. Jenkinson, M., Bannister, P., Brady, M., Smith, S.: Improved optimization for the robust and accurate linear registration and motion correction of brain images. Neuroimage **17**(2), 825–841 (2002)

9. Jenkinson, M., Smith, S.: A global optimisation method for robust affine registration of brain images. Med. Image Anal. **5**(2), 143–156 (2001)

10. Li, H., et al.: SuperDTI: Ultrafast DTI and fiber tractography with deep learning. Magn. Reson. Med. **86**(6), 3334–3347 (2021)

11. Liu, D., Wang, W., Wang, X., Wang, C., Pei, J., Chen, W.: Poststack seismic data denoising based on 3-D convolutional neural network. IEEE Trans. Geosci. Remote Sens. **58**(3), 1598–1629 (2019)

12. Ni, H., Kavcic, V., Zhu, T., Ekholm, S., Zhong, J.: Effects of number of diffusion gradient directions on derived diffusion tensor imaging indices in human brain. Am. J. Neuroradiol. **27**(8), 1776–1781 (2006)

13. Pajevic, S., Pierpaoli, C.: Color schemes to represent the orientation of anisotropic tissues from diffusion tensor data: application to white matter fiber tract mapping in the human brain. Magnet. Reson. Med.: Official J. Int. Soc. Magnetic Reson. Med. **42**(3), 526–540 (1999)

14. Sherbondy, A.J., Dougherty, R.F., Ben-Shachar, M., Napel, S., Wandell, B.A.: Contrack: finding the most likely pathways between brain regions using diffusion tractography. J. Vis. **8**(9), 15 (2008)

15. Tang, Y., et al.: Self-supervised pre-training of swin transformers for 3D medical image analysis. In: Proceedings of the IEEE/CVF Conference on Computer Vision and Pattern Recognition, pp. 20730–20740 (2022)

16. Tian, Q., et al.: DeepDTI: high-fidelity six-direction diffusion tensor imaging using deep learning. Neuroimage **219**, 117017 (2020)

17. Tian, Q., et al.: SDnDTI: self-supervised deep learning-based denoising for diffusion tensor MRI. Neuroimage **253**, 119033 (2022)

18. Van Essen, D.C., et al.: The human connectome project: a data acquisition perspective. Neuroimage **62**(4), 2222–2231 (2012)

19. Wasserthal, J., Neher, P., Maier-Hein, K.H.: TractSeg-fast and accurate white matter tract segmentation. Neuroimage **183**, 239–253 (2018)

20. Zhan, L., et al.: How does angular resolution affect diffusion imaging measures? Neuroimage **49**(2), 1357–1371 (2010)

FASSt: Filtering via Symmetric Autoencoder for Spherical Superficial White Matter Tractography

Yuan Li[1,2], Xinyu Nie[1,2], Yao Fu[3], and Yonggang Shi[1,2(✉)]

[1] Stevens Neuroimaging and Informatics Institute, Keck School of Medicine, University of Southern California (USC), Los Angeles, CA 90033, USA
yonggangs@usc.edu
[2] Ming Hsieh Department of Electrical and Computer Engineering, Viterbi School of Engineering, University of Southern California (USC), Los Angeles, CA 90089, USA
[3] Department of Computer and Data Sciences, Case School of Engineering, Case Western Reserve University (CWRU), Cleveland, OH 44106, USA

Abstract. Superficial white matter (SWM) plays an important role in functioning of the human brain, and it contains a large amount of cortico-cortical connections. However, the difficulties of generating complete and reliable U-fibers make SWM-related analysis lag behind relatively matured Deep white matter (DWM) analysis. With the aid of some newly proposed surface-based SWM tractography algorithms, we have developed a specialized SWM filtering method based on a symmetric variational autoencoder (VAE). In this work, we first demonstrate the advantage of the spherical representation and generate these spherical tracts using the triangular mesh and the registered spherical surface. We then introduce the Filtering via symmetric Autoencoder for Spherical Superficial White Matter tractography (FASSt) framework with a novel symmetric weights module to perform the filtering task in a latent space. We evaluate and compare our method with the state-of-the-art clustering-based method on diffusion MRI data from Human Connectome Project (HCP). The results show that our proposed method outperform these clustering methods and achieves excellent performance in groupwise consistency and topographic regularity.

Keywords: Tractography filtering · Superficial white matter · Spherical representation · Autoencoder

1 Introduction

Superficial white matter (SWM) contains short association fibers, which represent the cortico-cortical white matter connections in the human brain and are different from deep white matter (DWM) in both structure and function aspects [1]. The short association fibers or U-fibers, which connect the neighboring gyri, take up a large proportion of the cortico-cortical connectivity compared

This work is supported by the National Institute of Health (NIH) under grants R01EB022744, RF1AG077578, RF1AG056573, RF1AG064584, R21AG064776, U19AG078109, and P41EB015922.

M. Karaman et al. (Eds.): CDMRI 2023, LNCS 14328, pp. 129–139, 2023.
https://doi.org/10.1007/978-3-031-47292-3_12

to the long-range bundles [2]. Despite the importance of SWM for human brain mapping, SWM-related studies are often left behind compared to the DWM connectome research due to the high intrinsic variability of U-fibers' shape and localization across subjects [3]. Some previous works utilize the developed tools in DWM to perform the SWM analysis and generate anatomically meaningful U-fibers [4,5]. Recently, some surface-based methods have been proposed to generate more complete and reliable U-fibers [6,7]. With the aid of these methods, we can perform a more delicate SWM analysis by taking advantage of the underlying geometric information, and there is now a need for a more specific filtering algorithm for these U-fibers.

There are various methods for the tractography filtering task in DWM analysis. Clustering-based methods like QuickBundles [8] have been widely used to measure the similarity or distance of streamlines to remove outliers. Connectivity-based methods like [9,10] take advantage of the connectivity and geometric information and propose some regularity and consistency measurements to get rid of the outliers. Besides the above methods, some works like [11–15] use deep learning frameworks to handle the tractography clustering, segmentation and filtering task via different neural networks such as Autoencoder and CNN.

For SWM tractography filtering, some related works are using clustering-based approaches like [16,17] which use hierarchical clustering and fiber similarities to discard noisy fibers. [7] uses anatomical priors together with location information in pial surface mesh to filter out outliers. One challenge in SWM tractography filtering is that the complex nature of cortical folding and its variability across subjects undermines the reliability of the groupwise analysis for SWM. Additionally, the lack of ground truth makes evaluating the performance of filtering algorithm even harder.

In this work, we propose a SWM tractography filtering algorithm that uses the spherical representation of the U-fibers and variational autoencoder (VAE) [18] with a symmetric weights module to perform the filtering task in the latent space. Firstly, we convert the 3D Cartesian coordinate of U-fibers into the spherical representation using barycentric coordinates of the spherically registered cortical surface of each subject. Then, we propose a novel symmetric VAE to train the model using the generated spherical representation of U-fibers. At last, we demonstrate the advantage of our proposed symmetric weights module and evaluate the FASSt framework by calculating the groupwise consistency and topographic regularity of filtered fiber tracts.

2   Method

2.1   Spherical Representation of U-Fibers

Volume-based tractography methods such as [19,20] have been widely used to reconstruct fiber bundles in DWM but have limitations in reconstructing U-fibers. On the other hand, surface-based methods like [6,7] can reconstruct com-

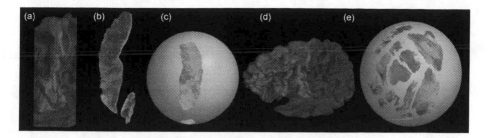

Fig. 1. 3D volume-based motor-sensory U-fibers (a,b) and its 2D spherical representation (c) for subject 101006 in HCP dataset. (e) is the spherical representation of the frontal lobe U-fibers (d) for the same subject.

plete and accurate U-fibers but still have limitations in generating false positive tracts.

To address these limitations, we propose a method that converts the 3D volume-based representation of U-fibers into a 2D spherical representation. A surface-based coordinate system that adapts to the variable cortical folding pattern of each individual allows for a more accurate localization of structure and functional features of the brain [21]. Unlike the registration of conventional tractography in 3D Cartesian coordinate, which relies heavily on the quality of diffusion and structural MRI, precomputed atlases, and may fail when encountering complex cortical anatomy with multiple crossing or poorly organized U-fibers, the spherical representation builds a bridge connecting different individuals' U-fibers. This enables a more accurate and robust groupwise comparison study in these highly variant and geometrically complicated cortical regions. We reconstruct spherical U-fibers using the barycentric coordinate generated by the surface-based tracking method and the registered subject sphere surface generated using FreeSurfer tools [22,23].

We denote the 3D volume-based representation of U-fibers for subject i to be $U_i = \{U_{ij}|j = 1, 2, ..., N\}$, barycentric coordinate for U_i to be $BaryCoord_i = \{BaryCoord_{ij}|j = 1, 2, ..., N\}$, and the spherical representation of U-fibers for subject i to be $S_i = \{S_{ij}|j = 1, 2, ..., N\}$, where N is the number of streamlines for this subject. U_{ij}, $BaryCoord_{ij}$ and S_{ij} represent a single streamline j of subject i in Cartesian coordinate, barycentric coordinate and spherical coordinate system. We first project U-fibers onto white matter surface to get barycentric coordinate $BaryCoord_i$ for U_i. Then, we get the spherical coordinate of the decimated white matter surface from the registered spherical surface. Finally, we compute the spherical representation of every point S_{ijk} (each streamline may contain a different number of points, k here ranges from one to the maximum number of points for S_{ij}) of a streamline S_{ij} using Eqs. 1:

$$
\begin{aligned}
S_{ijk} = &\ BaryCoord_{ijk}1 \times vertex1 + BaryCoord_{ijk}2 \times vertex2 \\
&+ BaryCoord_{ijk}3 \times vertex3
\end{aligned}
\tag{1}
$$

$Vertex1, 2, 3$ are the 3 vertices of a triangle which point k lies on the spherical registered white matter surface. $BaryCoord_{ijk}1, 2, 3$ are the barycentric coordinates of point k. Figure 1 shows an example of the 3D volume-based representation and spherical representation for motor-sensory U-fibers and frontal lobe U-fibers in HCP dataset.

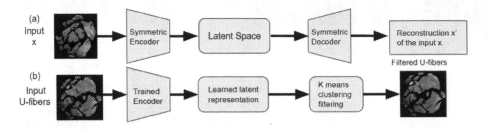

Fig. 2. FASSt has two compotents: (a) the scheme of a symmetric variational autoencoder, (b) filtering procedure for learned latent representation.

2.2 FASSt: Filtering via Symmetric Autoencoder for Spherical SWM Tractography

Recent works [16,17] have used a clustering-based approach for SWM tractography. In this work, we propose a filtering framework called FASSt (Filtering via symmetric Autoencoder for Spherical SWM tractography) that uses VAE and spherical coordinate. An autoencoder is a type of neural network that can learn efficient low-dimensional codings of high-dimensional data [24,25]. FASSt uses a VAE to learn the latent representation of U-fibers. VAE uses the variational Bayesian method to approximate the posterior distribution by reparameterizing the variational lower bound. The structure of the FASSt framework is shown in Fig. 2. The input x for the encoder is the resampled U-fibers (resampled into 128 points per streamline), and the learned latent representation is denoted as z. The encoder and decoder are represented by Eqs. 2 and 3:

$$Encoder(x) = q_\theta(z|x) \tag{2}$$

$$Decoder(x) = p_\phi(x|z) \tag{3}$$

The loss function for variational autoencoder consists of the log-likelihood with KL divergence term and other regularization terms, as shown in Eq. 4.

$$Loss(\theta, \phi) = \alpha \times D_{KL}(q_\theta(z|x)||p(z)) + \beta \times ELBO_{rec}(p_\phi(x|z)) \\ + \gamma \times ELBO_{ori}(p_\phi(x|z)) \tag{4}$$

D_{KL} represents the KL divergence, $ELBO_{rec}$ and $ELBO_{ori}$ represent the reconstruction loss and orientation loss, respectively (α, β, γ are the hyperparameters). The orientation loss is the normalized angle difference between two consecutive

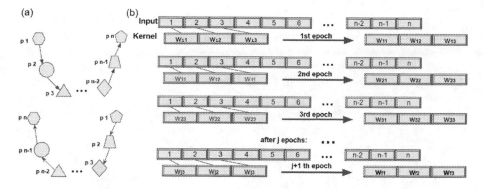

Fig. 3. Illustration of fiber orientation and symmetric weights module: (a) the same streamline is represent very differently when the orientation is not fixed, (b) symmetric weights module for $1D$ convolution kernel. We set kernel size to be 3 with *initial weights* $= (w_i1, w_i2, w_i3)$ and after $(j + 1)$th epoch (assume j is even number) we got the *final weights* $= (w_f1, w_f2, w_f3)$.

points along a streamline. This loss term aims to ensure the smoothness of the reconstructed streamlines. After learning the latent representations z, we can feed the unfiltered u-fibers to the encoder to obtain the corresponding latent representations of the bundle and perform filtering in the latent space using k-means clustering.

During training, the orientation of a streamline is not fixed and is very hard to be determined due to the complex shape of cortical folding pattern. As shown in Fig. 3 (a), this sequential representation of streamline can be ineffective to distinguish the same streamline but tracked in opposite directions [15] and may undermine the performance of our proposed method.

We propose a novel symmetric weights module (Fig. 3.b) to solve the U-fibers orientation problem. Firstly, we fixed our $1D$ convolution kernel size to be 3 for both the encoder and the decoder. Then, we set the weights kernel to be symmetric before each epoch i. We replace the third entry with the first entry of the kernel when i is an even number and replace the first entry with the third entry of the kernel when i is an odd number. The second entry is consistent through the whole process. Finally, we repeat this procedure until the stop criteria is satisfied and we get our final weights. Unlike the FiberMap feature descriptor in [15] which is computed by flipping and repeating the coordinates for every fiber in the whole tractography, this symmetric weights module utilizes the information from both directions along the training process without calculating additional fibers features.

2.3 Evaluation Method

Visual inspection for dense U-fibers like motor-sensory U-fibers is not always reliable. We adopt groupwise consistency [10] and topographic regularity [26] measurements to quantitatively evaluate our filtered results.

Groupwise consistency metric is used to demonstrate the effectiveness of the symmetric module of our proposed method. To compute groupwise consistency, we find the P closest streamlines for every subject in the reference set, and then we select Q subjects that are closest to the subject i for every streamline S_{ij}. Finally, we compute the mean distance between S_{ij} and the selected reference streamlines. We repeat the same procedure for every subject and compute the mean value of these distances, and use it as our groupwise consistency measurement across subjects. Smaller groupwise consistency value indicates better performance. In practice, we first resample the U-fibers to make sure that every streamline S_{ij} has the same number of points. Then, we subsample the U-fibers to make sure every subject have the same number of streamlines.

We use the proposed topographic regularity metric in [26] to compare our method with the 3D volume-based U-fibers filtering using FASSt (use the same structure of our proposed symmetric VAE but with 3D Cartesian coordinate), the QuickBundles filtering approach and simple filtering by minimum fiber length. To compute the topographic regularity, we first use classical multidimensional scaling to project the start and end points of the motor-sensory U-fibers onto a 2D plane, and then compute the Procrustes distance which measures the shape difference of start and end points as our topographic regularity measurement. A smaller topographic regularity value indicates a better result for a filtering method. As shown in Fig. 1 (b,c), there are U-fibers with disconnected parts, and it is difficult to separate their start points and end points because the closeness in Euclidean distance does not necessarily mean the closeness in the geodesic distance. Therefore, we extract the boundaries of the corresponding sulcal patches and then cluster these endpoints of U-fibers into start point set and end point set based on the smallest Euclidean distance of each point to the boundaries of the sulcal patch.

3   Experiment and Evaluation

We use 158 subjects from HCP dataset [27] in this study. The T1-weighted MRI images have resolution $= 0.7 \times 0.7 \times 0.7mm^3$. The diffusion MRI images have resolution $= 1.25 \times 1.25 \times 1.25mm^3$ with 18 b0 images and 90 volumes with b values $= 1000, 2000, 3000$. The tractography is generated using Probabilistic Tracking of U-fibers on the SWM Surface [6]. We randomly sampled 3 times at each triangle of the selected white matter triangle meshes. We set the step size to be 0.1mm with a maximum angle of steps at $10°C$. Our U-fiber tracking results contain approximately 10000 streamlines (at least 6000) per subject for both motor-sensory and frontal lobe regions.

After we get the SWM U-fibers, we then compute the corresponding spherical representation. The implementation details and parameter setting for the symmetric VAE training are as follows. For FASSt, the encoder contains six $1D$ convolution layers (32,64,128,256,512,1024,respectively). The latent representation length is 32 and latent vector is Gaussian distribution with mean value μ random sampled from $[0, 0.5]$ and standard deviation $\sigma = 0.5$. Decoder contains

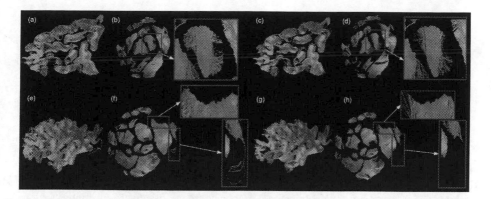

Fig. 4. Filtering results for subject 101006 and 114318 in HCP dataset. Subject 101006: unfiltered U-fibers for frontal lobe (a,b), filtering results for frontal lobe (c,d); Subject 114318: unfiltered U-fibers for frontal lobe (e,f), filtering results for frontal lobe (g,h), where (a,c,e,g) are the volume-based U-fibers and (b,d,f,h) are the converted spherical representation of the corresponding U-fibers.

six $1D$ convolution layers (1024,512,256,128,64,32, respectively). ReLU activations are used for both encoder's and decoder's convolution layers. Adam [28] optimizer is used with a learning rate 0.0008 for 500 epochs. Hyperparameters for loss function are fine-tuned as: $\alpha = 0.01, \beta = 1, \gamma = 0.0001$. We split the dataset to training set (100 subjects), validation set (18 subjects) and test set (40 subjects). Validation set is used to make sure that the trained model not only has minimum training loss but also has minimum validation loss. We use k-means clustering to generate clusters of the latent representation (k=50). We then discard some clusters containing very few streamlines and remove some streamlines that are far away from the center of the remaining clusters. In this step, we discard approximately 20 percent of the streamlines for the proposed method and all other methods to ensure a fair comparison. Finally, we obtain the filtered spherical U-fibers and transform them back into the 3D volume-based U-fibers. Figure 4 shows an example of our filtering algorithm, and we can observe obvious qualitative results of our filtering algorithm in the blue boxes (b,d,f,h).

To verify the effectiveness of our proposed symmetric weights module, we first train a model with same parameters and set up as mentioned above except that we do not fix the kernel weights to be symmetric at the beginning of each epoch. We then flip the orientation of every streamline and obtain the latent representation with symmetric weights module and without symmetric weights module for the frontal lobe and motor-sensory U-fibers, respectively. Finally, we calculate the Euclidean distance between the original and the flipped latent representation generated using our symmetric weights module and without using our symmetric weights module for each streamline. We then calculate the mean Euclidean distance for each subject in test set. As shown in Fig. 5, the mean Euclidean distance between the same streamline in opposite directions generated using proposed symmetric weights module is much smaller than the distance

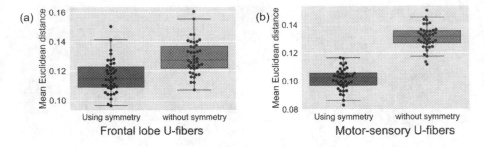

Fig. 5. Effectiveness demonstration of the symmetric weight module for test set in HCP dataset: the mean Euclidean distance calculated using the proposed symmetric weights module and without using symmetric weights module for frontal lobe U-fibers (a) and motor-sensory U-fibers (b).

computed without using symmetric weights module for both frontal lobe U-fibers (a) and motor-sensory U-fibers (b). This indicates that our proposed symmetric weights module is more robust in handling the variability of different orientations than conventional VAE.

We use groupwise consistency to demonstrate our filtering results and the important role of symmetric weights module. We use 100 subjects in HCP dataset as the reference set and calculate the groupwise consistency of the 2D spherical representations of U-fibers for 40 test subjects in the following manner. For each streamline, we select the closest 10 subjects in the reference set, and extract the 10 closest streamlines for each of the 10 subjects. We repeat this same procedure for every test subject. We calculate the groupwise consistency for both frontal lobe U-fibers and motor-sensory U-fibers. Figure 6 (a) shows that our proposed method has the smallest groupwise consistency value, indicating that our filtered frontal lobe U-fibers are more consistent across subjects. We observe that the mean value and variance of using symmetric weights module are slightly smaller than without using symmetric weights module in Fig. 6 (b). This could be explained by the fact that the motor-sensory U-fibers already has relative well organized U-fibers, and thus the impact of using symmetric weight module in terms of groupwise consistency is more subtle than the frontal lobe U-fibers.

We use the motor-sensory U-fibers in the test set to quantitatively measure the topographic regularity of FASSt and compare with the 3D volume-based U-fibers filtering using FASSt, the 2D spherical representation using FASSt but without symmetric weights module, the QuickBundles filtering approach and a simple filtering by minimum length method. Figure 7 shows that the performance of 3D volume-based U-fibers filtering using FASSt (group B) and QuickBundles (group F) are very close to each other, while our proposed method (group D) outperforms both of these methods. The reason why the performance of using symmetric module (group D) is only slightly better than without using symmetric weights module (group c) could be that the shape of motor-sensory U-fibers is more regular than other complex brain structures. This approach which we

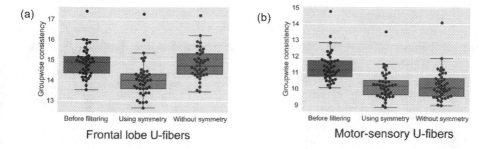

Fig. 6. Groupwise consistency comparison results: (a,b) show the box plots of group-wise consistency before any filtering, using FASSt and using FASSt but without symmetric module for frontal lobe U-fibers and motor-sensory U-fibers.

use 2D spherical representation and FASSt may already learn most of the useful features even without symmetric weights module. The similar performance of our proposed method (group D) and using FASSt with flipped U-fibers (group E) indicates that our proposed symmetric weights module is robust to the flipped U-fibers.

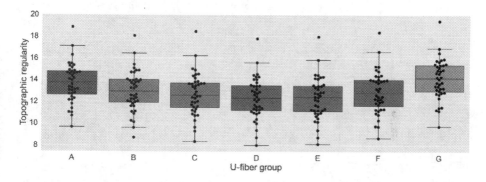

Fig. 7. Topographic regularity and groupwise consistency comparison results: this figure shows the box plot of topographic regularity of group A (blue: before any filtering), group B (orange: filtering using 3D volume-based data and FASSt), group C (green: filtering using 2D spherical representation and FASSt but without symmetric weights module), group D (red: filtering using 2D spherical representation and FASSt), group E (purple: same as C with flipped U-fibers), group F (brown: filtering using QuickBundles), group G (pink: filtering using minimum fiber length). (Color figure online)

4   Conclusion

Based on the spherical representation of U-fibers, we proposed in this work a novel U-fiber filtering method using VAEs with a symmetric weights module.

We evaluated our results via topographic regularity and groupwise consistency across subjects and demonstrated its excellent performance in comparison with state-of-the-art clustering methods. In future work, we will examine the impact of our U-fiber filtering method in the early detection of superficial white matter connectivity changes in clinical neuroimaging research. We also need to test our method in other brain regions with complex and variable SWM anatomy. Without a gold standard, incorporation of expertise from neurologist and radiologist may be valuable in assessing the validity of filtering results, which will be an important area of future work.

References

1. Kirilina, E., et al.: Superficial white matter imaging: contrast mechanisms and whole-brain in vivo mapping. Sci. Adv. **6**(41), eaaz9281 (2020)
2. Schüz, A., Braitenberg, V.: The human cortical white matter: quantitative aspects of cortico-cortical long-range connectivity. Cortical areas: Unity and diversity. 377–385 (2002)
3. Guevara, M., Guevara, P., Román, C., Mangin, J.F.: Superficial white matter: a review on the DMRI analysis methods and applications. Neuroimage **212**, 116673 (2020)
4. Catani, M., et al.: Short frontal lobe connections of the human brain. Cortex **48**(2), 273–291 (2012)
5. Vergani, F., et al.: White matter connections of the supplementary motor area in humans. J. Neurol. Neurosurg. Psych. **85**(12), 1377–1385 (2014)
6. Nie, X., Shi, Y.: Probabilistic tracking u-fiber on the superficial white matter surface. bioRxiv pp. 2022–05 (2022)
7. Shastin, D., et al.: Surface-based tracking for short association fibre tractography. Neuroimage **260**, 119423 (2022)
8. Garyfallidis, E., Brett, M., Correia, M.M., Williams, G.B., Nimmo-Smith, I.: Quickbundles, a method for tractography simplification. Front. Neurosci. **6**, 175 (2012)
9. Wang, J., Aydogan, D.B., Varma, R., Toga, A.W., Shi, Y.: Modeling topographic regularity in structural brain connectivity with application to tractogram filtering. Neuroimage **183**, 87–98 (2018)
10. Xia, Y., Shi, Y.: Groupwise track filtering via iterative message passing and pruning. Neuroimage **221**, 117147 (2020)
11. Legarreta, J.H., Petit, L., Rheault, F., Theaud, G., Lemaire, C., Descoteaux, M., Jodoin, P.M.: Filtering in tractography using autoencoders (FINTA). Med. Image Anal. **72**, 102126 (2021)
12. Li, B., et al.: Neuro4neuro: a neural network approach for neural tract segmentation using large-scale population-based diffusion imaging. Neuroimage **218**, 116993 (2020)
13. Bertò, G., et al.: Classifyber, a robust streamline-based linear classifier for white matter bundle segmentation. Neuroimage **224**, 117402 (2021)
14. Gupta, V., Thomopoulos, S.I., Corbin, C.K., Rashid, F., Thompson, P.M.: Fibernet 2.0: an automatic neural network based tool for clustering white matter fibers in the brain. In: 2018 IEEE 15th International Symposium on Biomedical Imaging (ISBI 2018), pp. 708–711. IEEE (2018)

15. Zhang, F., Karayumak, S.C., Hoffmann, N., Rathi, Y., Golby, A.J., O'Donnell, L.J.: Deep white matter analysis (DeepWMA): fast and consistent tractography segmentation. Med. Image Anal. **65**, 101761 (2020)
16. Román, C., et al.: Clustering of whole-brain white matter short association bundles using HARDI data. Front. Neuroinform. **11**, 73 (2017)
17. Mendoza, C., et al.: Enhanced automatic segmentation for superficial white matter fiber bundles for probabilistic tractography datasets. In: 2021 43rd Annual International Conference of the IEEE Engineering in Medicine & Biology Society (EMBC), pp. 3654–3658. IEEE (2021)
18. Kingma, D.P., Welling, M.: Auto-encoding variational bayes. arXiv preprint arXiv:1312.6114 (2013)
19. Tournier, J.D., Calamante, F., Connelly, A.: MRtrix: diffusion tractography in crossing fiber regions. Int. J. Imaging Syst. Technol. **22**(1), 53–66 (2012)
20. Aydogan, D.B., Shi, Y.: Parallel transport tractography. IEEE Trans. Med. Imaging **40**(2), 635–647 (2020)
21. Fischl, B., Sereno, M.I., Tootell, R.B., Dale, A.M.: High-resolution intersubject averaging and a coordinate system for the cortical surface. Hum. Brain Mapp. **8**(4), 272–284 (1999)
22. Dale, A.M., Fischl, B., Sereno, M.I.: Cortical surface-based analysis: I. segmentation and surface reconstruction. Neuroimage **9**(2), 179–194 (1999)
23. Fischl, B., Sereno, M.I., Dale, A.M.: Cortical surface-based analysis: II: inflation, flattening, and a surface-based coordinate system. Neuroimage **9**(2), 195–207 (1999)
24. Kramer, M.A.: Nonlinear principal component analysis using autoassociative neural networks. AIChE J. **37**(2), 233–243 (1991)
25. Hinton, G.E., Salakhutdinov, R.R.: Reducing the dimensionality of data with neural networks. Science **313**(5786), 504–507 (2006)
26. Aydogan, D.B., Shi, Y.: Tracking and validation techniques for topographically organized tractography. Neuroimage **181**, 64–84 (2018)
27. Van Essen, D.C., et al.: The human connectome project: a data acquisition perspective. Neuroimage **62**(4), 2222–2231 (2012)
28. Kingma, D.P., Ba, J.: Adam: A method for stochastic optimization. arXiv preprint arXiv:1412.6980 (2014)

Anisotropic Fanning Aware Low-Rank Tensor Approximation Based Tractography

Johannes Gruen[1,2] , Jonah Sieg[1], and Thomas Schultz[1,2(✉)]

[1] Institute for Computer Science, University of Bonn, Bonn, Germany
`schultz@cs.uni-bonn.de`
[2] Bonn-Aachen International Center for Information Technology, Bonn, Germany

Abstract. Low-rank higher-order tensor approximation has been used successfully to extract discrete directions for tractography from continuous fiber orientation density functions (fODFs). However, while it accounts for fiber crossings, it has so far ignored fanning, which has led to incomplete reconstructions. In this work, we integrate an anisotropic model of fanning based on the Bingham distribution into a recently proposed tractography method that performs low-rank approximation with an Unscented Kalman Filter. Our technical contributions include an initialization scheme for the new parameters, which is based on the Hessian of the low-rank approximation, pre-integration of the required convolution integrals to reduce the computational effort, and representation of the required 3D rotations with quaternions. Results on 12 subjects from the Human Connectome Project confirm that, in almost all considered tracts, our extended model significantly increases completeness of the reconstruction, at acceptable excess and additional computational cost. Its results are also more accurate than those from a simpler, isotropic fanning model that is based on Watson distributions.

Keywords: Fanning · Bingham distribution · Unscented Kalman filter

1 Introduction

Diffusion MRI tractography [11] permits the in-vivo reconstruction of white matter tracts in surgery planning or scientific studies. Spherical deconvolution is widely used to account for intra-voxel heterogeneity by estimating a continuous fiber orientation density function (fODF) in each voxel [7]. Representing fODFs

Funded by the Deutsche Forschungsgemeinschaft (DFG, German Research Foundation) - 422414649. Data were provided by the Human Connectome Project, WU-Minn Consortium (Principal Investigators: David Van Essen and Kamil Ugurbil; 1U54MH091657) funded by the 16 NIH Institutes and Centers that support the NIH Blueprint for Neuroscience Research; and by the McDonnell Center for Systems Neuroscience at Washington University.

as higher-order tensors and applying a low-rank approximation to these tensors has been shown to be a robust and efficient approach to estimating discrete tracking directions [1, 20].

However, while low-rank approximation accounts for fiber crossings, it ignores fiber fanning [21]. Consequently, even though recent work [9] has achieved promising results by performing low-rank approximation within the framework of Unscented Kalman Filter (UKF) based tractography [17], some fanning bundles were extracted incompletely when using single-region seeding strategies [9].

We address this limitation by explicitly modeling anisotropic fanning in the low-rank UKF with Bingham distributions [13]. This involves three main technical challenges: Firstly, initializing additional parameters in the UKF state. Section 3.2 solves this by observing that the Hessian matrix at the optimum of the low-rank approximation indicates the amount and direction of fanning. Secondly, the computational effort of convolving rank-one tensors with Bingham distributions. Section 3.3 solves this by pre-computing lookup tables for the corresponding integrals. Thirdly, maintaining a full 3D rotation per fiber compartment. Section 3.4 solves this with a quaternion-based representation. Results in Sect. 4 indicate that our extension reconstructs fanning bundles significantly more completely, at an acceptable additional computational cost.

2 Background and Related Work

2.1 Low-Rank Tensor Approximation Model

Constrained spherical deconvolution (CSD) computes the fiber orientation distribution function (fODF), a mapping from the sphere to $\mathbb{R}_+$ which captures the fraction of fibers in any direction [22]. One widely used strategy for estimating principal fiber orientations is to consider local fODF maxima. Our work builds on a variation of CSD, which represents the fODF as a symmetric higher-order tensor $\mathcal{T}$ and estimates r fiber directions via a rank-r approximation

$$\mathcal{T}^{(r)} = \sum_{i=1}^{r} \alpha_i \mathbf{v}_i^{\otimes l}, \tag{1}$$

where the scalar $\alpha_i \in \mathbb{R}_+$ denotes the volume fraction of the ith fiber, $\mathbf{v}_i \in \mathbb{S}^2$ its direction, and the superscript $\otimes l$ indicates an l-fold symmetric outer product, which turns the vector into an order-l tensor. Tensor order l corresponds to the truncation order when using spherical harmonics, while rank r is the number of fibers that are assumed to be present in a given fODF. The main benefit of this approach is that it can separate crossing fibers even if they are not distinct local maxima, which permits the use of lower orders and in turn improves numerical conditioning and computational effort [1]. Specifically, the angular resolution of fourth-order tensor approximation for crossing fibers has been shown to exceed order-eight fODFs with peak extraction [1]. To additionally capture information about anisotropic fanning, our current work increases the tensor order to $l = 6$, which parameterizes each fODF with 28 degrees of freedom.

2.2   Bingham Distribution

The Bingham distribution [3] is the spherical and antipodally symmetric ($f(\mathbf{x}) = f(-\mathbf{x})$) analogue to a two dimensional Gaussian distribution. It is given by the probability density function

$$f(\mathbf{x}; \mathbf{M}, \mathbf{Z}) := \frac{1}{N(\mathbf{Z})} \exp\left(\mathbf{x}^T \mathbf{M} \mathbf{Z} \mathbf{M}^T \mathbf{x}\right), \tag{2}$$

where $\mathbf{Z}$ is a diagonal matrix with decreasing entries $z_1 \geq z_2 \geq z_3$, $\mathbf{M} = (\mu_1, \mu_2, \mu_3)$ is an orthogonal matrix and $N(\mathbf{Z})$ denotes the hypergeometric function of matrix argument. Without loss of generality, we set $z_3 = 0$ and rename $\kappa = z_1, \beta = z_2$ to rewrite the density function as:

$$f(\mathbf{x}; \mu_1, \mu_2, \kappa, \beta) = \frac{1}{N(\kappa, \beta)} \exp\left(\kappa\langle\mu_1, \mathbf{x}\rangle^2 + \beta\langle\mu_2, \mathbf{x}\rangle^2\right). \tag{3}$$

Previously, Riffert et al. [18] fitted a mixture of Bingham distributions to the fODF to compute metrics such as peak spread and integral over peak. Kaden et al. [13] used it for Bayesian tractography. Our contribution combines the Bingham distribution with the low-rank model, and estimates the resulting parameters with a computationally efficient Unscented Kalman Filter.

2.3   Unscented Kalman Filter

The Kalman Filter is an algorithm that estimates a set of unknown variables, typically referred to as the state, from a series of noisy observations over time. The Unscented Kalman Filter (UKF) [12] is an extension that permits a non-linear relationship between the unknown variables and the measurements. It has first been used for tractography by Malcolm et al. [16,17], who treat the diffusion MR signal as consecutive measurements along a fiber, and the parameters of a mixture of diffusion tensors [16] or Watson distributions [17] as the unknown variables. Compared to independent estimation of model parameters at each location, this approach reduces the effects of measurement noise by combining local information with the history of previously encountered values. Consequently, it has been used for scientific studies [4,6] as well as neurosurgical planning [5].

Recent work has used the UKF to estimate the parameters of the low-rank model [9]. This variant of the UKF treats the fODFs instead of the raw diffusion MR signal as its measurements, which increases tracking accuracy while reducing computational cost, due to the much lower number of fODF parameters compared to diffusion-weighted volumes. A remaining limitation of that approach is that it does not account for fanning.

3   Material and Methods

We extend the previously described low-rank UKF [9] by modeling directional fanning with a Bingham distribution (Sect. 3.1). Implementing this requires solving problems related to initialization (Sect. 3.2), efficient evaluation of certain integrals (Sect. 3.3), and representing rigid body orientations within the UKF (Sect. 3.4). Section 3.5 describes the resulting tractography algorithm, while Sect. 3.6 reports the data and measures that we use for evaluation.

3.1 Low-Rank Model with Anisotropic Fanning

The higher-order tensor variant of CSD adapts the deconvolution so that it maps the single fiber response to a rank-one tensor [20]. Therefore, fanning can be incorporated by convolving the rank-1 kernel k with the Bingham distribution

$$h^{(r)} = \sum_{i=1}^{r} \alpha_i f\left(\cdot; \mu_1^{(i)}, \mu_2^{(i)}, \kappa^{(i)}, \beta^{(i)}\right) \star k, \tag{4}$$

where α_i denotes the volume fraction of the ith fiber in direction $\mu_1^{(i)} = \mathbf{v}_i$, $\kappa^{(i)}$ the concentration around it, i.e., the inverse to the amount of fanning. In case of anisotropic fanning, $\beta^{(i)} > 0$ indicates the additional fanning in direction $\mu_2^{(i)}$. For $\kappa^{(i)} \to \infty$ and $\beta^{(i)} = 0$, the Bingham distributions converge to delta peaks and the model (4) converges towards the original low-rank model (1).

3.2 Initialization via the Low-Rank Model

Since it is difficult to fit the model in Eq. (4) to data, we initialize the UKF based on the original low-rank approximation in Eq. (1). Firstly, we use the same main fiber directions, $\mu_1^{(i)} = \mathbf{v}_i$. Secondly, we initialize the fanning related parameters by observing that the rate at which the approximation error grows when rotating a given fiber direction away from its optimum depends on the amount of fanning: The lower the amount of fanning (the sharper the fODF peak), the more sensitive is the approximation error to the exact direction.

For each fiber, this information is captured in the second derivatives of the cost function with respect to its orientation, i.e., a 2×2 Hessian that can be computed in spherical coordinates; an equation for this is derived in [19]. There is a one-to-one mapping between the eigenvalues of that Hessian and corresponding values of κ and β. The eigenvector corresponding to the lower eigenvalue indicates the dominant fanning direction μ_2.

We pre-compute a lookup table for the values of κ and β, given the Hessian eigenvalues. To this end, we utilize the model (4) to generate single fiber fODFs for various combinations of κ and β values, and record the resulting eigenvalues. Figure 1 visualizes the mapping from eigenvalues to κ and β. Subfigure 1a shows that κ increases with the larger eigenvalue, indicating a higher concentration around the main fiber direction. Subfigure 1b shows how β depends on both eigenvalues. If they are the same, fanning is isotropic ($\beta = 0$), while for any given larger eigenvalue (EV1), β increases, indicating an increasingly elliptic fanning, as the smaller eigenvalues (EV2) decreases towards zero.

We apply this lookup table to multi-fiber voxels by computing the residual fODF for each fiber, i.e., subtracting out the remaining fibers, and normalizing

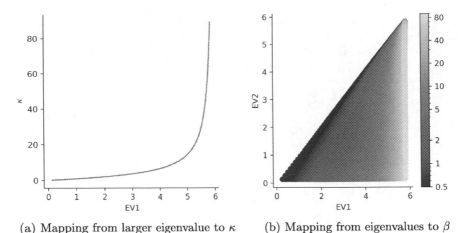

(a) Mapping from larger eigenvalue to κ (b) Mapping from eigenvalues to β

Fig. 1. Lookup tables for initializing κ and β based on Hessian eigenvalues. Left: κ is fully determined by the larger eigenvalue. Right: $\beta = 0$ if both eigenvalues are the same, but increases as EV2 decreases for a fixed value of EV1.

it such that $\alpha = 1$. After fixing all fiber directions and fanning parameters, we fit the volume fractions α_i in Eq. (4) with a non-negative least squares solver.

3.3   Pre-computing the Convolution

Equation (4) involves a convolution between a rank-1 kernel and a Bingham distribution. To compute it efficiently, we first split the Bingham distribution into a standard version and a rotation part. We rewrite

$$f\left(\mathbf{x}; \mathbf{M}, \mathbf{Z}\right) = \left(D\left(\vartheta, \psi, \omega\right) g\right)\left(\mathbf{x}; \kappa, \beta\right) = g\left(\mathbf{M}^{-1}\mathbf{x}; \kappa, \beta\right), \tag{5}$$

where $g\left(\mathbf{x}, \kappa, \beta\right) := \frac{1}{N(\mathbf{Z})} \exp\left(\kappa \mathbf{x}_3^2 + \beta \mathbf{x}_2^2\right)$ is a standard Bingham distribution in the canonical basis oriented towards the north pole and D is the zyz rotation matrix, which is defined as

$$\mathbf{M} = D\left(\vartheta, \psi, \omega\right) := R_z\left(\vartheta\right) R_y\left(\psi\right) R_z\left(\omega\right) \tag{6}$$

with rotation matrices $R_y(\alpha)$ and $R_z(\alpha)$ that rotate by α around the y or z axis, respectively. This decomposition is a significant simplification, because we can now pre-compute the convolution between the standard Bingham distribution and the kernel, and apply the rotation afterwards.

As it is standard practice in CSD [22], we perform the convolution on the sphere using spherical and rotational harmonics. A rotational harmonics representation of the rank-1 kernel has been computed previously [20]. Unfortunately, no closed form solution is available for the spherical harmonics coefficients of the Bingham distribution. Therefore, we pre-compute them numerically, for the relevant range of $\kappa \in \{2.1, 2.2, \ldots, 89\}$ and $\beta \in \{0, 0.1, \ldots, \kappa - 2\}$.

3.4   Representing Rotations with Quaternions

Unlike previous UKF-based tractography methods, our model requires a full three-dimensional rotation per fiber to account not just for the fiber direction, but also for the direction of its anisotropic spread. Unit quaternions are a popular representation of rotations, since they overcome limitations of Euler angles, such as gimbal lock. However, integrating them into a UKF is non trivial, since their normalization leads to dependencies within the state [14]. We overcome the problem by utilizing a homeomorphism between quaternions and $\mathbb{R}^3$. We discuss it below, but refer the reader to Bernal-Polo et al. [2] for a more detailed discussion of quaternions and their integration into the UKF. Introductions to UKF-based tractography are also available in the literature [9, 16, 17].

Given a quaternion $q = [q_w, q_x, q_y, q_z] \in \mathbb{H}$, we define a homeomorphism

$$\phi : \mathbb{S}^3 \to \left\{ \mathbf{e} \in \mathbb{R}^3 : \|\mathbf{e}\| \leq 4 \right\} \tag{7}$$

$$q \mapsto 4 \frac{q_{1:}}{1 + q_0} \tag{8}$$

to the so-called Modified Rodrigues Parameters [26] and, vice versa,

$$\phi^{-1} : \left\{ \mathbf{e} \in \mathbb{R}^3 : \|\mathbf{e}\| \leq 4 \right\} \to \mathbb{S}^3 \tag{9}$$

$$\mathbf{e} \mapsto \frac{1}{16 + \|\mathbf{e}\|^2} \left(16 - \|\mathbf{e}\|^2, 8\mathbf{e} \right). \tag{10}$$

In a close neighborhood of the identity quaternion, these charts behave like the identity transformation between the imaginary part of quaternions and $\mathbb{R}^3$. For a given mean quaternion $\bar{q}$ of a quaternions set $\{q_i\}_i$, we define a mapping which first maps each q_i by the conjugated mean quaternion, pushes it to $\mathbb{R}^3$, $\phi_{\bar{q}}(q_i) := \phi(\bar{q}^\star \star q_i) = \mathbf{e}$, where $\bar{q}^\star = [\bar{q}_w, -\bar{q}_x, -\bar{q}_y, -\bar{q}_z]$ denotes the conjugated quaternion, pulls it back and rotates it back via $\phi_{\bar{q}}^{-1}(\mathbf{e}) := \bar{q} \star \phi^{-1}(\mathbf{e})$. Assuming that the quaternions are highly concentrated around the mean quaternion, the embedding resembles the distribution of quaternions closely.

With these preliminaries, we set up the UKF as illustrated in Fig. 2. For simplicity, our implementation updates the parameters for each fiber separately. The state at point t is defined by the parameters of a single Bingham distribution in the embedded space, $X_t := \{\alpha, \kappa, \beta, \mathbf{e}_1, \mathbf{e}_2, \mathbf{e}_3\}$. The embedding is fully determined by the quaternion q_t and the covariance is denoted by P_t.

We create sigma points to capture the distribution of the covariance around the current mean. We use the sigma points to calculate a chart update q_{t+1}, by taking a weighted mean with weights w_i and pulling the embedded part back into quaternion space. With the new chart we perform a chart transition. Afterwards, we follow the standard UKF update scheme: Firstly, calculate the weighted mean of the sigma points, evaluate our model for all sigma points and take the corresponding weighted mean. Secondly, calculate the covariance P_{xx} of the sigma points, the covariance of the evaluation P_{zz} and the cross correlation P_{xy}. This information is then used to calculate the Kalman gain K and correct the current state X_t dependent on the difference between the expected measurement $\bar{Z}$ and the fODF z as well as the covariance.

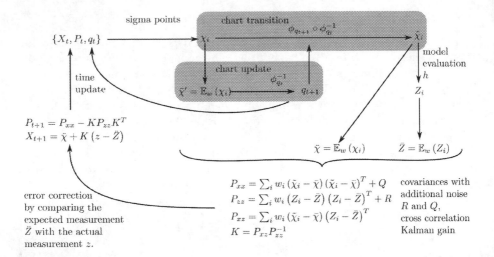

Fig. 2. Schematic representation of an update step in our filter. In addition to the steps of a traditional UKF, we update the charts by first calculating the weighted mean of the sigma points and pulling it back into quaternion space. With the new quaternion, we perform a chart transition.

3.5   Probabilistic Streamline-Based Tractography

For a given seed point, we initialize the UKF as discussed in Sect. 3.2. We perform streamline integration with second-order Runge-Kutta: At the jth point of the streamline, we update the UKF, select the Bingham distribution whose main direction is closest by angle to the current tracking direction, and draw a direction from that Bingham distribution via rejection sampling. We use that direction for a tentative half-step, again update the UKF and perform rejection sampling. Finally, we reach point $(j+1)$ by taking a full step from point j in that new direction. This process is iteratively conducted until a stopping criterion is reached. We stop the integration if the white matter density drops below 0.4 or if we cannot find any valid direction within 60°. In our experiments, white matter density was derived from multi-shell dMRI data [1]. In single-shell data, it could be obtained from a tissue type segmentation of a T_1 image [9].

3.6   Data and Evaluation

It is the goal of our work to modify the UKF so that it more completely reconstructs fanning bundles from seeds in a single region. We evaluate this on 12 subjects from the Human Connectome Project (HCP) [23] for which reference tractographies have been published as part of TractSeg [25]. They are based on a segmented and manually refined whole-brain tractography. We evaluate reconstructions of these tracts from seed points that we obtain by intersecting the reference bundles with a plane, and picking the initial tracking direction that is closest to the reference fiber's tangent at the seed point. We estimate fODFs using data from all three b shells that are available in the HCP data [1].

We use a step size of 0.5 mm and seed three times at each seed point leading to three times as many streamlines as the reference before filtering. Due to the probabilistic nature of our method, we perform density filtering to remove single outliers. Since diffusion MRI tractography is known to create false positive streamlines [15], we also apply filtering based on inclusion and exclusion regions similar to the ones described by Wakana et al. [24]. We place those regions manually in a single subject, and transfer them to the remaining ones via linear registration. Any streamline that does not intersect with all inclusion regions or intersects with an exclusion region is removed entirely.

To make the comparison against the previously described low-rank UKF [9] more direct, we set its tensor order to 6. We also evaluate the benefit of modeling *anisotropic* fanning by implementing a variant of our approach that uses an isotropic Watson distribution, and could be seen as an extension of the previously proposed Watson UKF [17]. For this model and for the Bingham UKF, we conducted a visually guided parameter optimization and finalized $Q = \{\alpha = 0.05, \kappa = 0.05, v_1 = 0.02, v_2 = 0.02, v_3 = 0.02\}$ and $R = 0.02$ for the Watson UKF and $Q = \{\alpha = 0.01, \kappa = 0.1, \beta = 0.1, e_1 = 0.005, e_2 = 0.005, e_3 = 0.005\}$ and $R = 0.02$ for the Bingham UKF. For all models we set the fiber rank to 2.

We judge the completeness and excess of all tractographies based on distances between points on the reference tracts, and the generated ones. Specifically, we employ the 95% quantile $\chi^{95\%}$ of the directed Hausdorff distance

$$h(A, B) := \chi^{95\%} \left\{ \min_{b \in B} \|a - b\| : a \in A \right\}, \tag{11}$$

where A and B denote point sets [10]. Intuitively, if the 95% quantile of $h(A, B) = d$, then 95% of the vertices of A are within distance d from some point of B. This measure is not symmetric. Thus, setting A to the reference tractography and B to the reconstruction penalizes false negatives (it scores completeness), while switching the arguments penalizes false positives (it scores the excess).

4   Results

Figure 3 presents a qualitative comparison of the reconstruction of the Cingulum (CG) in an example subject. In comparison to the low-rank UKF, both the Watson UKF and the Bingham UKF result in a more complete reconstruction of the parahippocampal part a). Moreover, compared to the Watson UKF, the Bingham UKF achieves a more complete reconstruction of fibers entering the anterior cingulate cortex b). Similar trends are observed in Fig. 4 for the reconstruction of the cortospinal tract (CST). The Bingham UKF successfully reconstructs a majority of the lateral fibers, while both the low-rank UKF and the Watson UKF are missing some parts of the fanning. However, at the bottom, it also leads to visually more pronounced outliers.

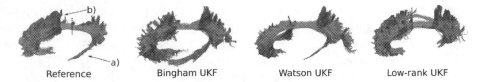

Fig. 3. Reconstructions of the left Cingulum, black dotted line denotes seed points. The Bingham UKF permits the most complete reconstruction. The low-rank UKF misses the parahippocampal part a), while the Watson UKF misses fibers towards the anterior cingulate cortex b).

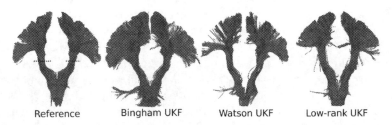

Fig. 4. The corticospinal tract (CST), black dotted line denotes seed points. The Bingham UKF leads to the highest streamline density in the lateral fanning.

We quantify these results by evaluating directed Hausdorff distances. The upper part of Fig. 5 shows distances from the reference to the reconstruction. In 6 out of 7 tracts, the Bingham UKF exhibits the lowest median, indicating the most complete reconstructions. The lower part measures distances from the reconstruction to the reference, so that low values indicate low excess. In 6 out of 7 tracts, the Bingham UKF leads to a lower median than the low-rank UKF.

To statistically assess the differences between the proposed methods, we conducted non-parametric Friedman rank tests [8]. An asterisk denotes significant differences at significance level of $p < 0.007$, due to Bonferroni correction. In 6 out of 7 tracts, we found significant differences in the completeness of reconstruction. In 4 out of 7 tracts, significant differences were observed for the excess.

Generation of 1000 CST streamlines took 92.5 s for the Bingham UKF, 85.2 s for the Watson UKF, and 58.1 s for the low-rank UKF on a single core of a 3.3 GHz CPU.

5   Discussion

Visual results suggest that modeling fanning increases completeness, but leads to a slightly higher number of outliers. It is unsurprising that permitting stronger deviations from the main fiber direction increases the risk of getting diverted into neighboring tracts. On the other hand, quantitative results indicate that, in most tracts, excess is reduced in addition to the increased completeness. This can be explained by the use of the 95% Hausdorff metric, which ignores a small number of outliers, and has been selected because any probabilistic tractography

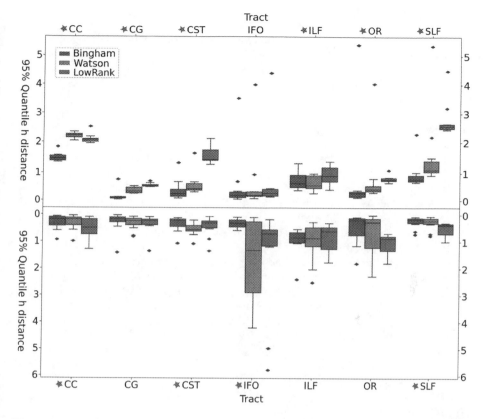

Fig. 5. Top: 95% quantile of the directed Hausdorff distance from reference to reconstruction. Median distances are lowest, indicating highest completeness, for Bingham UKF in all tracts except the ILF. In 6 out of 7 tracts, differences are significant (asterisks). Bottom: Directed distance from reconstruction to reference. In most tracts, even these distances decrease when modeling fanning, indicating improved specificity in addition to the higher completeness.

will generate a certain number of outliers. In any case, we consider the outliers in our results to be acceptable given the clear improvement in completeness.

Even though our experiments successfully used the same parameter values in all HCP subjects, the exact values that are specified above might have to be adjusted to achieve optimal results for dMRI data with very different characteristics, such as single-shell clinical data.

6   Conclusion

We developed a new algorithm for probabilistic tractography that incorporates anisotropic fanning into the recently described low-rank UKF. We demonstrated that, in almost all bundles, this results in more complete reconstructions, while keeping the excess at an acceptable level. Our proposed technical solutions for initialization, convolution, and representation of rotations contribute to maintaining acceptable computational efficiency. Our code is available as part of the bonndit package: https://github.com/MedVisBonn/bonndit

References

1. Ankele, M., Lim, L.H., Groeschel, S., Schultz, T.: Versatile, robust, and efficient tractography with constrained higher-order tensor fODFs. Int. J. Comput. Assisted Radiol. Surg. **12**(8), 1257–1270 (2017). https://doi.org/10.1007/s11548-017-1593-6
2. Bernal-Polo, P., Martínez-Barberá, H.: Kalman filtering for attitude estimation with quaternions and concepts from manifold theory. Sensors **19**(1), 149 (2019). https://doi.org/10.3390/s19010149
3. Bingham, C.: An antipodally symmetric distribution on the sphere. Ann. Stat. **2**(6), 1201–1225 (1974). https://doi.org/10.1214/aos/1176342874
4. Chen, Z., et al.: Corticospinal tract modeling for neurosurgical planning by tracking through regions of peritumoral edema and crossing fibers using two-tensor unscented Kalman filter tractography. Int. J. Comput. Assisted Radiol. Surg. **11**(8), 1475–1486 (2016). https://doi.org/10.1007/s11548-015-1344-5
5. Cheng, G., Salehian, H., Forder, J.R., Vemuri, B.C.: Tractography from HARDI using an intrinsic unscented Kalman filter. IEEE Trans. Med. Imaging **34**(1), 298–305 (2015). https://doi.org/10.1109/TMI.2014.2355138
6. Dalamagkas, K., et al.: Individual variations of the human corticospinal tract and its hand-related motor fibers using diffusion MRI tractography. Brain Imaging Behav. **14**, 696–714 (2020). https://doi.org/10.1007/s11682-018-0006-y
7. Dell'Acqua, F., Tournier, J.D.: Modelling white matter with spherical deconvolution: how and why? NMR Biomed. **32**(4) (2018). https://doi.org/10.1002/nbm.3945
8. Friedman, M.: The use of ranks to avoid the assumption of normality implicit in the analysis of variance. J. Am. Stat. Assoc. **32**(200), 675–701 (1937). https://doi.org/10.1080/01621459.1937.10503522
9. Grün, J., Gröschel, S., Schultz, T.: Spatially regularized low-rank tensor approximation for accurate and fast tractography. NeuroImage **271** (2023). https://doi.org/10.1016/j.neuroimage.2023.120004
10. Huttenlocher, D., Klanderman, G., Rucklidge, W.: Comparing images using the Hausdorff distance. IEEE Trans. Pattern Anal. Mach. Intell. **15**(9), 850–863 (1993). https://doi.org/10.1109/34.232073
11. Jeurissen, B., Descoteaux, M., Mori, S., Leemans, A.: Diffusion MRI fiber tractography of the brain. NMR Biomed. **32**(4), e3785 (2019). https://doi.org/10.1002/nbm.3785
12. Julier, S., Uhlmann, J.: Unscented filtering and nonlinear estimation. Proc. IEEE **92**(3), 401–422 (2004). https://doi.org/10.1109/JPROC.2003.823141
13. Kaden, E., Knösche, T.R., Anwander, A.: Parametric spherical deconvolution: inferring anatomical connectivity using diffusion MR imaging. Neuroimage **37**(2), 474–488 (2007). https://doi.org/10.1016/j.neuroimage.2007.05.012
14. Kraft, E.: A quaternion-based unscented Kalman filter for orientation tracking. In: Proceedings of the Sixth International Conference of Information Fusion, vol. 1, pp. 47–54 (2003). https://doi.org/10.1109/ICIF.2003.177425

15. Maier-Hein, K., et al.: The challenge of mapping the human connectome based on diffusion tractography. Nat. Commun. **8**, 1349 (2017). https://doi.org/10.1038/s41467-017-01285-x

16. Malcolm, J., Shenton, M., Rathi, Y.: Filtered multitensor tractography. IEEE Trans. Med. Imaging **29**, 1664–1675 (2010). https://doi.org/10.1109/TMI.2010.2048121

17. Malcolm, J.G., Michailovich, O., Bouix, S., Westin, C.F., Shenton, M.E., Rathi, Y.: A filtered approach to neural tractography using the Watson directional function. Med. Image Anal. **14**(1), 58–69 (2010). https://doi.org/10.1016/j.media.2009.10.003

18. Riffert, T.W., Schreiber, J., Anwander, A., Knösche, T.R.: Beyond fractional anisotropy: extraction of bundle-specific structural metrics from crossing fiber models. Neuroimage **100**, 176–191 (2014). https://doi.org/10.1016/j.neuroimage.2014.06.015

19. Schultz, T., Kindlmann, G.: A maximum enhancing higher-order tensor glyph. Comput. Graph. Forum **29**(3), 1143–1152 (2010). https://doi.org/10.1111/j.1467-8659.2009.01675.x

20. Schultz, T., Seidel, H.P.: Estimating crossing fibers: a tensor decomposition approach. IEEE Trans. Vis. Comput. Graph. **14**(6), 1635–1642 (2008). https://doi.org/10.1109/TVCG.2008.128

21. Sotiropoulos, S.N., Behrens, T.E., Jbabdi, S.: Ball and rackets: inferring fiber fanning from diffusion-weighted MRI. Neuroimage **60**(2), 1412–1425 (2012). https://doi.org/10.1016/j.neuroimage.2012.01.056

22. Tournier, J.D., Calamante, F., Connelly, A.: Robust determination of the fibre orientation distribution in diffusion MRI: non-negativity constrained super-resolved spherical deconvolution. Neuroimage **35**(4), 1459–1472 (2007). https://doi.org/10.1016/j.neuroimage.2007.02.016

23. Van Essen, D.C., Smith, S.M., Barch, D.M., Behrens, T.E., Yacoub, E., Ugurbil, K.: The WU-Minn human connectome project: an overview. Neuroimage **80**, 62–79 (2013). https://doi.org/10.1016/j.neuroimage.2013.05.041

24. Wakana, S., et al.: Reproducibility of quantitative tractography methods applied to cerebral white matter. Neuroimage **36**, 630–644 (2007). https://doi.org/10.1016/j.neuroimage.2007.02.049

25. Wasserthal, J., Neher, P., Maier-Hein, K.H.: TractSeg - fast and accurate white matter tract segmentation. Neuroimage **183**, 239–253 (2018). https://doi.org/10.1016/j.neuroimage.2018.07.070

26. Wiener, T.F.: Theoretical analysis of gimballess inertial reference equipment using delta-modulated instruments. Ph.D. thesis, Massachusetts Institute of Technology (1962)

BundleCleaner: Unsupervised Denoising and Subsampling of Diffusion MRI-Derived Tractography Data

Yixue Feng[1(✉)], Bramsh Q. Chandio[1], Julio E. Villalón-Reina[1],
Sophia I. Thomopoulos[1], Himanshu Joshi[3], Gauthami Nair[3], Anand A. Joshi[2],
Ganesan Venkatasubramanian[3], John P. John[3], and Paul M. Thompson[1]

[1] Imaging Genetics Center, Keck School of Medicine, University of Southern
California, Marina del Rey, CA, USA
yixuefen@usc.edu
[2] Signal and Image Processing Institute, Ming Hseih department of Electrical and
Computer Engineering, University of Southern California, Los Angeles, CA, USA

[3] Multimodal Brain Image Analysis Laboratory, Translational Psychiatry
Laboratory, National Institute of Mental Health and Neuro Sciences, Bengaluru,
Karnataka, India

Abstract. We present *BundleCleaner*, an unsupervised multi-step
framework that can filter, denoise and subsample bundles derived from
diffusion MRI-based whole-brain tractography. Our approach considers
both the global bundle structure and local streamline-wise features. We
apply *BundleCleaner* to bundles generated from single-shell diffusion
MRI data in an independent clinical sample of older adults from India
using probabilistic tractography and the resulting 'cleaned' bundles can
better align with the atlas bundles with reduced overreach. In a downstream tractometry analysis, we show that the cleaned bundles, represented with less than 20% of the original set of points, can robustly localize along-tract microstructural differences between 32 healthy controls
and 34 participants with Alzheimer's disease ranging in age from 55 to
84 years old. Our approach can help reduce memory burden and improving computational efficiency when working with tractography data, and
shows promise for large-scale multi-site tractometry.

Keywords: dMRI · tractography · denoising · tractometry

1 Introduction

Diffusion MRI (dMRI) allows us to examine white matter (WM) pathways in
vivo and structural connectivity in the brain. Whole-brain tractograms (WBT),

This work was supported by the NIH grants RF1AG057892 and R01AG060610,
the Department of Science and Technology, Govt. of India, grant nos. DST-
SR/CSI/73/2011 (G); DST-SR/CSI/70/2011 (G); and DST/CSRI/2017/249 (G).

composed of streamlines, are used to model WM and can be generated using various tractography methods. However, inferring the geometry of neural pathways indirectly from water diffusion using tractography has limitations in practice: bundles in the brain often intersect, making it difficult to accurately infer long-range projections and determine termination points near the cortex [13]. In practice, WBTs may typically contain over 1 million streamlines per subject, including a high percentage of false positives (streamlines that do not represent true anatomy), and these also persist after bundle segmentation [27]. This can cause problems for downstream tasks, such as tractometry, where detected signals can be due to artifacts instead of true group differences. Many factors can affect the proportion of false positive streamlines - the quality of the incoming imaging data, the spatial and angular resolution of the acquisition, the reconstruction method for estimating the fiber orientation distribution function (fODF), and the tractography and bundle segmentation methods. This is a major concern when the dMRI data is collected in the single-shell low b-value regime. Large-scale multi-site neuroimaging consortia, such as ENIGMA, have processed large amounts of dMRI data most of it being single shell, low b-value data ($b = 600-1200\,\text{s/mm}^2$) [12,16]. Post-hoc bundle filtering techniques become especially important in this context to reduce the proportion of spurious fibers that can affect the statistical analysis at the bundle level.

To address post-hoc filtering of tractography data, deep learning methods, such as an autoencoder [17] and geometric models [4], have been proposed for direct streamline filtering in WBT, where each streamline used in training is labeled as either valid or invalid. However, these models evaluate each streamline independently and are trained on reliably labeled WBTs instead of segmented bundles. Spurious fibers can occur in bundles where the streamline itself is structurally plausible but deviates from the overall fiber bundle shape. FiberNeat [6], filters streamlines in the low-dimensional embedding space using dimensionality reduction methods t-SNE and UMAP. The input to t-SNE and UMAP is a pairwise streamline distance matrix, which does not take into account the full structure of streamlines. In addition, both methods apply filtering by assigning 'valid' and 'invalid' labels to each streamline. However, when a streamline deviates only slightly or contains structural anomalies in a local neighborhood, it could potentially be addressed with smoothing instead of removal.

In this study, we propose *BundleCleaner*, an unsupervised multi-step framework to filter, denoise, and subsample bundles using both point cloud and streamline-based methods. This approach considers the global bundle structure as well as local streamline-wise features, and can be applied to bundles of varying shapes and proportions of erroneous streamlines. We applied *BundleCleaner* to bundles generated from single-shell dMRI in an independent cohort from India, and show that the cleaned bundles can improve alignment with the atlas bundles and reduce overreach. In a downstream tractometry analysis conducted using BUndle ANalytics (BUAN) [5], the cleaned bundles, represented with less than 20% of the original set of points, can robustly identify group differences in microstructural measures, with reduced memory burden and increased efficiency.

Code for *BundleCleaner* is publicly available at https://github.com/wendyfyx/BundleCleaner (Fig. 1).

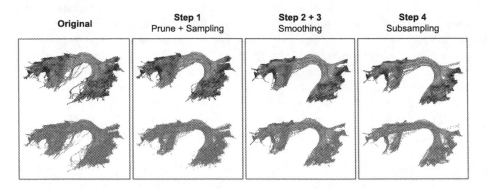

Fig. 1. Here we show the 4 steps of *BundleCleaner* (see Sect. 3) on an example arcuate fasciculus bundle, comprised of two steps of pruning (steps 1 and 4), and two steps of smoothing (steps 2 and 3). Bundles are shown with both streamline (top) and point-cloud (bottom) representations.

2   Materials

We analyzed 3T single-shell dMRI data of the human brain in a pilot sample of 66 participants from the NIMHANS (National Institute of Mental Health and Neuro Sciences) cohort (mean age: 67.1 ± 7.4 years; 26F/40M). Thirty-four participants were diagnosed with Alzheimer's disease (AD) and 32 were cognitively normal controls (CN). The dMRI data were acquired using a single-shell, diffusion weighted echo-planar imaging sequence (TR = 7441 ms, TA = 630 s, TE = 85 ms, voxel size: $2 \times 2 \times 2$ mm^3, 64 slices, flip angle = 90°, field of view (224 mm)^2). The protocol consisted of 64 diffusion-weighted ($b = 1000$ s/mm^2), and 1 $b = 0$ s/mm^2 volume, where diffusion weighting was encoded along 64 independent orientations. Transverse sections of 2 mm thickness were acquired parallel to the anterior commissure-posterior commissure line. dMRI data were preprocessed to correct for artifacts with DIPY [9] and FSL [14], including noise [18], Gibbs ringing [15,19], susceptibility induced distortions [1], eddy currents [2] and bias field inhomogeneity [26]. Whole brain tractograms were generated using constrained spherical convolution (CSD) [25] and probabilistic particle filtering tracking (PFT) [11] with the following parameters - 8 seeds per voxel generated from the WM mask, step size of 0.2 mm, angular threshold of 30°, and the continuous map stopping criterion. RecoBundles [10] was used to extract 38 bundles for each subject in both native and MNI (Montreal Neurological Institute) space using a standard atlas in MNI ICBM 2009c space [28].

3   BundleCleaner

In developing filtering or denoising methods for white matter bundles, it would be beneficial to design methods that consider the sequential information within streamlines. However, some outliers can be determined by local features instead of all points on the streamline. It is important to design a smoothing algorithm that considers the global cohesion of the bundle as well as the smoothness of local streamlines. With this in mind, *BundleCleaner* contains the following steps:

- **Step 1:** Streamline resampling and pruning using QuickBundles [8]
- **Step 2:** Point cloud-based smoothing using Laplacian regularization [24];
- **Step 3:** Streamline-based smoothing using the Savitzky-Golay filter [23];
- **Step 4:** Streamline pruning and subsampling using QuickBundles (similar to step 1)

Methods used in each step are detailed below, and all parameters are listed with their default values in Table 1.

Table 1. *BundleCleaner* parameters and their default values.

		BundleCleaner Parameters	Default Values
Step 1	S_r	Resampling rate	0.5
	D	Distance threshold	5 mm
	C	Minimum cluster size	*auto* (median cluster size)
Step 2	α	Smoothing parameter	100
	K	Number of neighbors for triangulation	50
Step 3	W	Smoothing window size	5
Step 4	D'	Distance threshold	5 mm
	C'	Minimum cluster size	*auto* (median cluster size)
	S	Subsampling rate	0.5 (depend on use cases)

3.1   Step 1: Streamline Resampling and Pruning

Each bundle is initially loaded as a set of streamlines or 3D point sequences. Step size during fiber tracking is often smaller than the voxel size to produce more accurate streamlines [20], resulting in a dense representation. However, we can adequately represent streamlines with fewer points than those directly produced by the tractography [21]. To reduce the computational cost in Step 2 while maintaining the fixed step size, each streamline in the bundle is first resampled to $S_r\%$ points equally spaced along the line.

As bundles with high levels of noise can be difficult to smooth directly, an initial step of pruning is applied using QuickBundles [8], an efficient streamline-based clustering method with a distance threshold D. The minimum direct flip distance (MDF) is used in clustering since streamlines are flip-invariant. Starting from one initial cluster, each streamline whose MDF distance to any existing centroid is smaller than D is added to the corresponding cluster; otherwise, it is added to a new cluster. A preset distance threshold ensures that clustering applied to different bundles yields a consistent density of streamlines. A minimum cluster size C is a hyperparameter such that any cluster smaller than C is pruned in this step. The bundle is then converted into a point-set representation for point cloud based Laplacian smoothing in the next step.

3.2   Step 2: Point Cloud-Based Smoothing

While point clouds remove the sequential information in the streamlines, a smoothing operation applied to this representation can still consider a local neighborhood of points, which can be from neighboring streamlines. We assume that streamlines close to each other in a local neighborhood are more likely to share similar trajectories. One of the most common smoothing techniques, Laplacian regularization, has been successfully applied to point clouds [29,30] using graph-based methods. In our work, the Laplacian matrix for a point cloud is computed using the approach described in Sharp and Crane (2022) [24]. They presented a robust replacement of the commonly used cotangent Laplacian by creating a "tufted cover" for non-manifold edges and vertices and flipping edges to build Delaunay triangulations so that the Laplacian always has non-negative edge weights. This method can proves to be extremely robust, considering that point cloud triangulations contain many non-manifold edges or vertices. This approach is very fast, averaging 2.2 minutes to smooth point clouds ranging from 20K to 300K points in our analysis. Reconstruction of an example arcuate fasciculus with the Laplacian eigenfunctions is shown in Fig. 2.

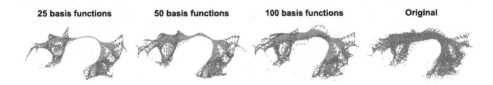

Fig. 2. Point cloud based reconstruction of an arcuate fasciculus bundle with 25, 50, 100 eigenfunctions of the Laplacian matrix. The original bundle contains 62,253 points. The Laplacian matrix was built with a $K = 100$ sized neighborhood.

To compute the point cloud Laplacian, a local neighborhood of K neighbors is used to build a triangulation. The point cloud is then smoothed using Laplacian regularization,

$$C(X_s) = ||X_s - X_o||^2 + \alpha||\Delta X_s||^2$$
$$C(Y_s) = ||Y_s - Y_o||^2 + \alpha||\Delta Y_s||^2$$
$$C(Z_s) = ||Z_s - Z_o||^2 + \alpha||\Delta Z_s||^2$$

where X_s, Y_s, Z_s are the smoothed coordinates, X_o, Y_o, Z_o are the original coordinates, and α is the smoothing parameter, where larger α applies heavier smoothing. Each objective is smoothed using the conjugate gradient method.

3.3   Step 3: Streamline Based Smoothing

The bundle as a point cloud is reshaped back to streamlines for an additional step of smoothing using the Savitzky-Golay filter [23], an approach often used in 1D signal smoothing. The filter is applied to each streamline independently with second order polynomial and preset window size W, in case the denoised point clouds from Step 2 contain jagged edges. In addition, Sarwar *et al.* [22] have described the oscillating trajectories present in streamlines generated from probabilistic CSD as a *rippling* artifact, not present in those generated from deterministic methods. Higher W is therefore recommended for bundles generated from probabilistic tractography to produce smoother streamlines (Fig. 3).

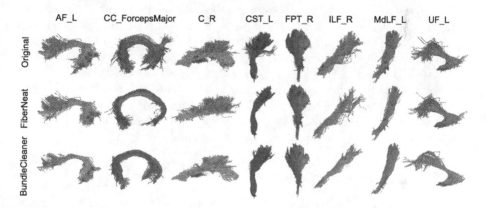

Fig. 3. Examples of the 8 original, bundles cleaned with both *FiberNeat* and *BundleCleaner*. Parameters for *BundleCleaner* are described in Sect. 4

3.4   Step 4: Streamline-Based Subsampling

For efficient downstream tractometry analysis, BundleCleaner contains a step of streamline-based subsampling with a pruning procedure similar to Step 1 described in Sect. 3.1. While smoothing in Steps 2 and 3 would ideally remove and denoise spurious streamlines, extracted bundles can still contain streamlines that deviate locally in their orientation and curvature when the local neighborhood used in constructing the point-cloud Laplacian are all from outlier streamlines. The extra pruning in this step will further remove streamlines that could not be denoised. QuickBundles based clustering is applied to the output of Step 3 with the two parameters - distance threshold D' and minimum cluster size C' - similar to Step 1. The difference is that random subsampling is applied to each remaining cluster larger than the C' with a subsampling rate of S.

4   Experiments and Results

We applied *BundleCleaner* to 8 bundles from each of the 66 participants - the left *arcuate fasciculus* (AF_L), corpus callosum's *forceps major* (CC_ForcepsMajor), right *cingulum* (C_R), left *corticospinal tract* (CST_L), right *fronto-pontine tract* (FPT_R), left *inferior longitudinal fasciculus* (ILF_R), right *middle longitudinal fasciculus* (MdLF_L), and left *uncinate fasciculus* (UF_L) - with the following adjustments to the default parameters: $S_r = 0.3$, $\alpha = 150$, $K = 50$, $W = 10$ and $S = 0.75$. *FiberNeat* [6] was applied to all 528 bundles using UMAP method[1] for comparison with *BundleCleaner*. For bundles with fewer than 50 streamlines, only streamline smoothing (Step 3) is applied. Given the high levels of irregularities and the large number of streamlines generated from probabilistic fiber tracking [22], we resampled 30% points ($S_r = 0.3$) resulting in roughly 100 points per streamline on average, and apply more aggressive smoothing with a larger α, K and W value. After resampling, the distance between consecutive points in all streamlines remains smaller than the voxel size of 2 mm, to ensure that dMRI-derived microstructural measures can be adequately mapped to points on the bundle. With $S_r = 0.3$, $S = 0.75$, and default pruning parameters, the resulting cleaned bundles contain 17% of points and 60% of streamlines compared to the original bundles on average. These downsampling factors were applied to allow data reduction to enable more efficient downstream analysis while retaining fidelity to the original tract geometry. To evaluate the results of *Bundle-Cleaner*, we conducted a quantitative shape comparison and demonstrated how the cleaned bundle performs in localizing group differences in diffusion tensor imaging (DTI) measures using the BUAN tractometry pipeline.

4.1   Evaluation: Shape Comparison

We conduct shape comparison of the bundles before and after cleaning by computing two shape metrics - bundle *overlap* and *overreach* with respect to the atlas bundle [7]. We first voxelize each bundle b and the atlas bundle a using the same reference density map image into binary mask images V_b and V_a, and then computing the fraction of non-zero voxels in V_b overlapping and overreaching with respect to V_a.

$$\text{Overlap}(V_b, V_a) = \frac{|V_b \cap V_a|}{|V_b|} \quad \text{and} \quad \text{Overreach}(V_b, V_a) = \frac{|V_b \setminus V_a|}{|V_b|}$$

Figure 4 plots the two metrics for 8 bundle types before and after cleaning. We can see increased bundle overlap and decrease overreach scores in both cleaning methods, with *BundleCleaner* outperforming *FiberNeat* in particularly in C_R, CST_L and FPT_R in both metrics. These results indicate that *Bundle-Cleaner*, by incorporating smoothing can improve bundle alignment with the atlas bundles.

---

[1] *FiberNeat* implementation is available at https://github.com/BramshQamar/FiberNeat/tree/main.

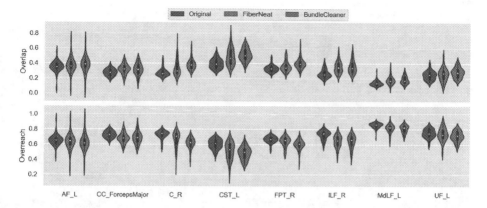

Fig. 4. Shape comparison of the original, *FiberNeat* and *BundleCleaner* bundles, using metrics - bundle overlap and coverage with the atlas bundle defined in Sect. 4.1.

4.2 Evaluation: Tractometry

BUAN, using all points in a bundle, applies linear mixed models (LMM) to localize group differences along its length. We ran the BUAN tractometry pipeline on the aforementioned bundles from 32 CN versus 34 AD subjects, and investigated group differences in four DTI measures - fractional anisotropy (FA), mean diffusivity (MD), axial diffusivity (AD) and radial diffusivity (RD) - before and after cleaning. Fixed effect terms in the LMM include age and sex in addition to diagnostic group, and the random effect term is the subjects to account for multiple points belonging to the same subject in a segment being analyzed. False discovery rate (FDR) correction was applied to each bundle to account for multiple comparisons along the 100 tract segments.

Here, we show results for one of the DTI measures for each of six bundles in Fig. 5 - plots on the left in each box show the negative logarithm of *p*-values on the *y*-axis, segments along the length of the bundle on the *x*-axis, along with FDR corrected threshold marked with a blue horizontal line. The standardized β values from the LMM are mapped onto the atlas bundle to indicate effect sizes on the right side of the box. The detected group differences are largely similar between the original and cleaned bundles - except for the end regions where shape variability is higher, and signals are less resistant to noise, notably in RD in segments 80–100 of the right cingulum. Other statistically significant differences after FDR correction are detected in the cleaned bundles - segment 10–30 in the right fronto-pontine tract, segment 20–40 in the left middle longitudinal fasciculus, and segment 30–40 in the left uncinate fasciculus - with similar if not larger effect size as reflected by the standardized β, despite using less than 20% of the points. This shows that *BundleCleaner* can robustly detect group differences in DTI measures while reducing the size of intermediate files by half.

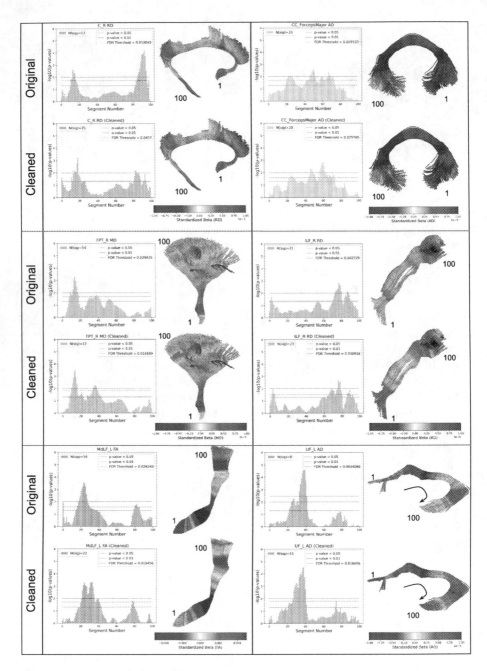

Fig. 5. BUAN results for six bundles before and after cleaning. For each bundle, a bar plot of the $-\log(p)$ with the FDR threshold marked in blue (left), and standardized β mapped along 100 segments of the bundle (right) are shown. The group difference is shown between CN and AD participants. The number of segments with p-values that pass the FDR threshold is also marked in the bar plot. (Color figure online)

5   Discussion

BundleCleaner parameters can be tuned to directly control global and local streamline smoothness as well as remove false positive. While a higher α value in Step 2 will produce "tighter" bundles with more cohesion, a higher W value in Step 3 will produce streamlines with smoother trajectories; a higher minimum cluster size C/C' in Step 1 and 4 will apply more aggressive streamline filtering. However, the approach relies on some degree of 'correctness' and overall cohesion of bundles produced by segmentation. In the case where a large proportion of the bundles is anatomically implausible, it is difficult for *BundleCleaner* to prune out such fibers. Figure 6 shows two examples in one CC_ForcepsMajor, where cleaning was too aggressive, and one UF_L bundle where it fails to remove a significant portion of false positive streamlines. We can reduce C/C' for less aggressive pruning in the first case, and increase them or manually clean bundles using QuickBundles [8] for the second case. In addition, when cleaning bundles such as the corticospinal tract (CST) - where the lateral branch is likely to be removed - or bundles with large diameter but with sparse streamlines such as the corpus callosum, we recommend reducing C/C' for a more lenient threshold.

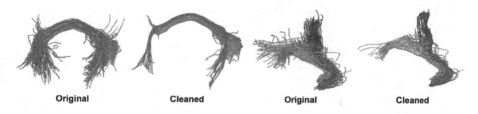

Original   **Cleaned**   **Original**   **Cleaned**

Fig. 6. Two failure cases in one corpus callosum forceps major bundle and one left uncinate fasciculus bundle.

While each step of *BundleCleaner* can be run independently of the others, some parameters can be tuned together to provide better synergies. Resampling and pruning in step 1 ensure that fewer points are passed on to step 2, since point cloud-based smoothing is the most time-consuming step of this process. Lower S_r in Step 1 is therefore recommended for bundles generated with small tractography step sizes. Another effect of a lower resampling rate S_r is that it is more likely for points from neighboring streamlines to be considered in Laplacian smoothing, and this can be more directly tuned with K in Step 2. Lower S_r and higher K apply global bundle-based smoothing over a larger neighborhood while reducing the runtime in Step 2, and may be helpful for bigger and denser bundles.

BundleCleaner was developed with application to tractometry in mind and we take inspiration from smoothing operations in voxel-based morphometry prior to statistical analysis to increase the signal-to-noise (SNR) ratio [3]. While smoothing can remove finer fiber structure when applied to tractography data, we believe that the manually adjustable parameters in our method can offer value in practice for different use cases, such as tractometry and structural connectivity analysis. In our future work, we plan to test our approach on multi-shell datasets with higher b-values, streamlines generated with deterministic tractography, and bundles segmented from different atlases, to make corresponding parameter recommendations.

6   Conclusion

BundleCleaner is an unsupervised multi-step framework that can filter, denoise, and subsample bundles using both point cloud and streamline-based methods, which considers the global bundle structure as well as local streamline-wise features. This approach is efficient, can denoise bundles with more than 1 million points, reduce bundle overreach and improve alignment with the atlas bundles used in segmentation, making it easier to work with unwieldy tractography data in downstream analysis. We show in a downstream tractometry analysis that cleaned bundles represented with less than 20% of the points compared to the original bundles can robustly localize microstructural differences between CN and AD subjects in an Indian cohort, while significantly reducing memory burden. This approach shows promise for large-scale multi-site tractometry analysis. *BundleCleaner* parameters are intuitive to tune and can be applied to any fiber bundles with varying noise levels.

References

1. Andersson, J.L., Skare, S., Ashburner, J.: How to correct susceptibility distortions in spin-echo echo-planar images: application to diffusion tensor imaging. Neuroimage **20**(2), 870–888 (2003). https://doi.org/10.1016/S1053-8119(03)00336-7
2. Andersson, J.L., Sotiropoulos, S.N.: An integrated approach to correction for off-resonance effects and subject movement in diffusion MR imaging. Neuroimage **125**, 1063–1078 (2016). https://doi.org/10.1016/j.neuroimage.2015.10.019
3. Ashburner, J., Friston, K.J.: Voxel-based morphometry-the methods. Neuroimage **11**(6), 805–821 (2000). https://doi.org/10.1006/nimg.2000.0582
4. Astolfi, P., et al.: Tractogram filtering of anatomically non-plausible fibers with geometric deep learning (2020). arXiv:2003.11013
5. Chandio, B.Q., et al.: Bundle analytics, a computational framework for investigating the shapes and profiles of brain pathways across populations. Sci. Rep. **10**(1), 17149 (2020). https://doi.org/10.1038/s41598-020-74054-4
6. Chandio, B.Q., et al.: FiberNeat: unsupervised streamline clustering and white matter tract filtering in latent space. Preprint, Neuroscience (2021). https://doi.org/10.1101/2021.10.26.465991
7. Côté, M.A., et al.: Tractometer: towards validation of tractography pipelines. Med. Image Anal. **17**(7), 844–857 (2013). https://doi.org/10.1016/j.media.2013.03.009
8. Garyfallidis, E., et al.: QuickBundles, a method for tractography simplification. Front. Neurosci. **6**, 175 (2012). https://doi.org/10.3389/fnins.2012.00175
9. Garyfallidis, E., et al.: Dipy, a library for the analysis of diffusion MRI data. Front. Neuroinform. **8** (2014). https://doi.org/10.3389/fninf.2014.00008

10. Garyfallidis, E., et al.: Recognition of white matter bundles using local and global streamline-based registration and clustering. Neuroimage **170**, 283–295 (2018). https://doi.org/10.1016/j.neuroimage.2017.07.015

11. Girard, G., Whittingstall, K., Deriche, R., Descoteaux, M.: Towards quantitative connectivity analysis: reducing tractography biases. Neuroimage **98**, 266–278 (2014). https://doi.org/10.1016/j.neuroimage.2014.04.074

12. Hatton, S.N., et al.: White matter abnormalities across different epilepsy syndromes in adults: an ENIGMA-Epilepsy study. Brain **143**(8), 2454–2473 (2020). https://doi.org/10.1093/brain/awaa200

13. Jbabdi, S., Johansen-Berg, H.: Tractography: where do we go from here? Brain Connectivity **1**(3), 169–183 (2011). https://doi.org/10.1089/brain.2011.0033

14. Jenkinson, M., Beckmann, C.F., Behrens, T.E., Woolrich, M.W., Smith, S.M.: FSL. NeuroImage **62**(2), 782–790 (2012). https://doi.org/10.1016/j.neuroimage.2011.09.015

15. Kellner, E., et al.: Gibbs-ringing artifact removal based on local subvoxel-shifts: Gibbs-ringing artifact removal. Magn. Reson. Med. **76**(5), 1574–1581 (2016). https://doi.org/10.1002/mrm.26054

16. Koshiyama, D., et al.: White matter microstructural alterations across four major psychiatric disorders: mega-analysis study in 2937 individuals. Mol. Psychiatry **25**(4), 883–895 (2020). https://doi.org/10.1038/s41380-019-0553-7

17. Legarreta, J.H., et al.: Filtering in tractography using autoencoders (FINTA). Med. Image Anal. **72**, 102126 (2021). https://doi.org/10.1016/j.media.2021.102126

18. Manjón, J.V., et al.: Diffusion weighted image denoising using overcomplete local PCA. PLoS ONE **8**(9), e73021 (2013). https://doi.org/10.1371/journal.pone.0073021

19. Neto Henriques, R.: Advanced methods for diffusion MRI data analysis and their application to the healthy ageing brain (2017). https://doi.org/10.17863/CAM.29356

20. O'Donnell, L.J., Westin, C.F.: An introduction to diffusion tensor image analysis. Neurosurg. Clinics North America **22**(2), 185–196, viii (2011). https://doi.org/10.1016/j.nec.2010.12.004

21. Presseau, C., et al.: A new compression format for fiber tracking datasets. Neuroimage **109**, 73–83 (2015). https://doi.org/10.1016/j.neuroimage.2014.12.058

22. Sarwar, T., et al.: Mapping connectomes with diffusion MRI: deterministic or probabilistic tractography? Magn. Reson. Med. **81**(2), 1368–1384 (2019). https://doi.org/10.1002/mrm.27471

23. Savitzky, A., Golay, M.J.E.: Smoothing and differentiation of data by simplified least squares procedures. Anal. Chem. **36**(8), 1627–1639 (1964). https://doi.org/10.1021/ac60214a047

24. Sharp, N., Crane, K.: A Laplacian for nonmanifold triangle meshes. Comput. Graph. Forum **39**(5), 69–80 (2020). https://doi.org/10.1111/cgf.14069

25. Tournier, J.D., Calamante, F., Connelly, A.: Robust determination of the fibre orientation distribution in diffusion MRI. Neuroimage **35**(4), 1459–1472 (2007). https://doi.org/10.1016/j.neuroimage.2007.02.016

26. Tournier, J.D., et al.: MRtrix3: a fast, flexible and open software framework for medical image processing and visualisation. Neuroimage **202**, 116137 (2019). https://doi.org/10.1016/j.neuroimage.2019.116137

27. Wasserthal, J., Neher, P., Maier-Hein, K.H.: TractSeg - fast and accurate white matter tract segmentation. Neuroimage **183**, 239–253 (2018). https://doi.org/10.1016/j.neuroimage.2018.07.070

28. Yeh, F.C., et al.: Population-averaged atlas of the macroscale human structural connectome and its network topology. Neuroimage **178**, 57–68 (2018). https://doi.org/10.1016/j.neuroimage.2018.05.027
29. Zeng, J., et al.: 3D point cloud denoising using graph Laplacian regularization of a low dimensional manifold model (2019). arXiv:1803.07252 [cs]
30. Zhang, S., et al.: Hypergraph spectral analysis and processing in 3D point cloud. IEEE Trans. Image Process. **30**, 1193–1206 (2021). https://doi.org/10.1109/TIP.2020.3042088

A Deep Network for Explainable Prediction of Non-imaging Phenotypes Using Anatomical Multi-view Data

Yuxiang Wei[1,2], Yuqian Chen[2,3], Tengfei Xue[2,3], Leo Zekelman[2],
Nikos Makris[2], Yogesh Rathi[2], Weidong Cai[3], Fan Zhang[2],
and Lauren J. O'Donnell[2(✉)]

[1] University of Electronic Science and Technology of China, Chengdu, China
[2] Harvard Medical School, Boston, USA
odonnell@bwh.harvard.edu
[3] University of Sydney, Sydney, Australia

Abstract. Large datasets often contain multiple distinct feature sets, or views, that offer complementary information that can be exploited by multi-view learning methods to improve results. We investigate anatomical-multi-view data, where each brain anatomical structure is described with multiple feature sets. In particular, we focus on sets of white matter microstructure and connectivity features from diffusion MRI, as well as sets of gray matter area and thickness features from structural MRI. We investigate machine learning methodology that applies multi-view approaches to improve the prediction of non-imaging phenotypes, including demographics (age), motor (strength), and cognition (picture vocabulary). We present an explainable multi-view network (EMV-Net) that can use different anatomical views to improve prediction performance. In this network, each individual anatomical view is processed by a view-specific feature extractor and the extracted information from each view is fused using a learnable weight. This is followed by a wavelet-transform-based module to obtain complementary information across views which is then applied to calibrate the view-specific information. Additionally, the calibrator produces an attention-based calibration score to indicate anatomical structures' importance for interpretation. In the experiments, we demonstrate that the proposed EMV-Net significantly outperforms several state-of-the-art methods designed for non-imaging phenotype prediction based on the Human Connectome Project (HCP) Young Adult dataset with 1065 individuals. EMV-Net significantly outperforms compared methods for predicting age, strength, and picture vocabulary. Specifically, our approach specifically decreases the Mean Absolute Error (MAE) for age prediction by at least 0.24 years and improves the correlation coefficient for predicting the other two phenotypes by at least 0.13. Our interpretation results show that for different views, fractional anisotropy of white matter diffusion measures and the surface thickness of gray matter measures are generally more important.

Keywords: dMRI · Brain White Matter · Multi-view Learning

© The Author(s), under exclusive license to Springer Nature Switzerland AG 2023
M. Karaman et al. (Eds.): CDMRI 2023, LNCS 14328, pp. 165–176, 2023.
https://doi.org/10.1007/978-3-031-47292-3_15

1   Introduction

The brain's anatomical structures are crucial in neurological function and neurodevelopment. Magnetic resonance imaging (MRI) enables quantitative analysis of the brain's structural properties. Structural MRI can measure the macroscopic morphometry of the cortical and subcortical structures in gray matter (GM), while diffusion MRI (dMRI) tractography [1] can assess white matter (WM) connectivity and is used to extract quantitative microstructure measures such as fractional anisotropy (FA) and mean diffusivity (MD) [35]. To uncover links between brain structure and non-imaging phenotypes including demographics or behavioral traits, a popular avenue of research is machine learning (ML) for the prediction of non-imaging phenotypes. For example, ML models including ridge regression [14] and a lightweight deep-learning model named SFCN (Simple Fully Convolutional Network) [21] have been applied to predict age from structural MRI data. Recently, a study [3] compared eight ML-based models for predicting age and cognitive functions using diffusion MRI connectivity data. They found that a multilayer perceptron network (MLP) achieved the best results.

Although the above methods successfully predicted different phenotypic traits, they did not specifically investigate the multi-view nature of MRI data. In this paper, we investigate multi-view learning [9,25,31,32], to handle a specific type of multi-view data *where the same set of objects (samples) is described by several distinct feature sets* [9]. This type of data occurs naturally in dMRI, where various microstructure models can offer unique information (multiple views) [2] to describe each anatomical connection. Similarly, parcellated structural MRI data can contain multiple views of each anatomical structure, such as the thickness and surface area of each cortical region. For clarity, we call this type of data *anatomical-multi-view data*, where multiple feature sets describe each anatomical structure of the brain's WM or GM. Most related work in multi-view learning for neuroimaging has focused on multimodal images or sets of extracted image features from multimodal imagery [3,10,18,24,26,33]. Thus, there has been limited focus on the development of methods specifically for anatomical-multi-view data derived from MRI.

Challenges for learning from anatomical-multi-view data include: (1) challenge to leverage the underlying feature patterns of each anatomical view, (2) challenge to fuse information from other views while learning view-specific features, (3) challenge to simultaneously interpret each view and each structure's importance for different learning tasks, and (4) challenge to develop a general model that can perform well across different input datasets and tasks. In response to these challenges, we introduce an Explainable Multi-View Network (EMV-Net). This network effectively captures unique information from each view across all anatomical structures and synthesizes multiple feature sets for specific structures. The network learns weights that quantify the importance of each view in the prediction process. In the parlance of a recent review of explainable artificial intelligence, this is considered to be a global (dataset-level) explanation [27]. The network is evaluated using two different inputs (white matter and gray matter anatomy) and on three different prediction tasks. Overall, the network significantly outperforms compared state-of-the-art methods in predicting different human phenotypes.

2   Methodology

2.1   Dataset

All data used in this paper is derived from the "1200 Subjects Data Release" dataset from the Human Connectome Project (HCP) Young Adult Study [30].

White Matter Feature Dataset. The dataset of diffusion measure included in this paper is derived from the "1200 Subjects Data Release" dataset from the Human Connectome Project (HCP) Young Adult Study [30]. The data processing pipeline for tractography and tract parcellation is described in [37]. Briefly, to compute whole-brain tractography, the two-tensor unscented Kalman filter (UKF) [12] method is employed via the ukftractography package of SlicerDMRI [36]. Following a recursive estimation order, the UKF method fits a mixture model of two tensors to the diffusion data while tracking fibers. The first tensor models the traced tract, and the second tensor models fibers crossing the tract. UKF tractography is highly consistent across ages, health conditions, and image acquisitions [34,38]. Afterward, tractography parcellation is performed based on a neuroanatomist-curated WM atlas [19,20,38]. The tractography parcellation contains 947 clusters after discarding expert-defined false positive clusters as in [38].

For each cluster, we compute a total of 22 features, where each feature is considered as a view of the cluster. We compute the FA and Trace from tensor 1 (FA1 and Trace1) and tensor 2 (FA2 and Trace2) at each streamline point and record their max, min, median, mean, and variance across all fiber points within the cluster. Therefore, given each tensor $t \in \{t_1, t_2\}$, we extract statistical measure $s \in \{max, min, median, mean, var\}$ of the diffusion parameters $p \in \{FA, Trace\}$, resulting in 20 features per cluster. We also compute the total number of points and streamlines in each cluster.

Gray Matter Feature Dataset. FreeSurfer [4] is a widely-used tool for automated segmentation and parcellation of brain regions from MRI. The dataset used in this study was generated by the Free-surfer 5.3.0-HCP pipeline and contains 68 cortical regions' measurements. For each region, the dataset offers two sets of features: the surface area that reflects the density of neurons in the anatomical region and the surface thickness that relates to the integrity of cortical tissue [16]. These features are considered two anatomical views in our study.

Predicting Tasks. We choose three popular brain-based prediction tasks: age [21] (demographic), strength [28] (motor), and picture vocabulary [3] (cognition).

2.2 Methods

The proposed method consists of three key steps. Firstly, individual feature extractors with sparse self-attention are applied to each view, effectively capturing underlying patterns. Secondly, a cross-view calibration technique is introduced to fuse features across views while retaining view-specific information. This calibration involves using a wavelet transform to uncover latent feature patterns in the frequency domain, enhancing the calibration process. Lastly, our network enables interpretable explanations of the importance of each anatomical view and brain region. By assigning learnable weights to each view before fusion, the calibrator provides insights into the importance of individual views. Additionally, the decomposed information from the wavelet transform in the calibrator is transformed into attention scores, which reveal the importance of different anatomical regions. The overall architecture of the proposed EMV-Net is shown in Fig. 1, which is used to extract view-specific features, and the cross-view calibrator (Fig. 2) that is proposed fuse features from multiple views to calibrate view-specific information and provide interpretability for each view and anatomical region's importance.

Single-View Backbone. For each view, we apply a feature extractor with a commonly used backbone, largely following the design of [5]. In brief, the backbone has three repeated modules (placed sequentially), where each module utilizes a patch embedding unit for downsampling, then extracts information with multiple sparse CMT (convolution-meets-transformer) blocks (S-CMT Blocks). By combining convolution and self-attention outputs, the backbone can extract both local and global information. Note that unlike [5], we use 1.5-Entmax [22] for self-attention, which can sparsely select the most relevant anatomical structures from each view.

Attentive Multiview Network. As shown in Fig. 1, we employ a multi-view-multi-net architecture like [26]. However, instead of only performing one fusion as in [26], we fuse information across views in each repeated module using the explainable cross-view calibrator, which enables view-specific feature extractors to learn complementary and relevant information from other views. After the last calibrator, feature maps from all views are concatenated, flattened, and then used to perform predictions. The overall architecture of the cross-view calibrator is shown in Fig. 2. It contains three steps: (1) highlighting relevant features from multiple views and fusing them, (2) performing wavelet transform on the fused features, and (3) sigmoid normalization to produce scores from the processed information and then applying the scores to calibrate view-specific features.

View-Specific Feature Highlighting and Fusion. Since different anatomical views can have different contributions to different tasks, applying flexible weights to the views could boost performance while granting interpretability. To highlight more

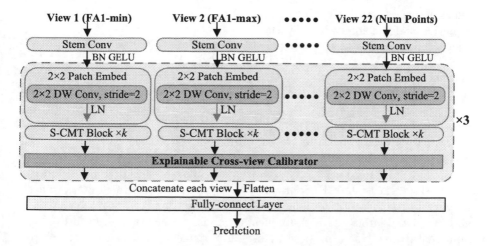

Fig. 1. The overall architecture of the proposed explainable attentive multi-view network. Each view is processed with a single view backbone. Note that "Stem Conv" denotes an initial convolution applied to each view, BN denotes batch normalization, GELU is the activation function, "Patch Embed" is the patch embedding process that is commonly used in Transformer, and DW Conv is the depth-wise convolution with stride 2.

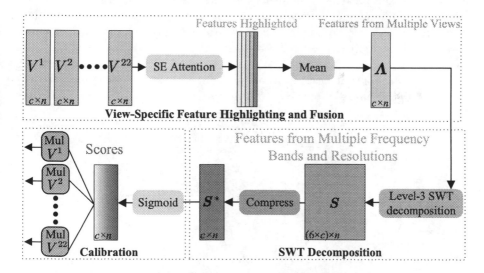

Fig. 2. The proposed explainable cross-view calibrator. n and c denote the feature length (number of features from all clusters) and the number of channels. Here SE Attention is the squeeze-and-excitation attention, and Mul denotes the matrix multiplication between the scores and the original input from each view. The features from each view are firstly selected by the SE attention, then fused through mean. The fused features are subsequently decomposed by SWT and then used to generate a score with Sigmoid activation.

relevant views, we first apply a squeeze-and-excitation module [8] to produce attention weights for each view (here we treat each view as a channel) and assign these weights to the corresponding view features. The selected features from each view are then fused by averaging, which produces the fused feature map Λ.

SWT Decomposition for Improved Calibration Across Views. He wavelet transform [13] is a popular signal processing tool and has been applied to denoise or extract features from structural and diffusion MRI [6,7,15]. It can reveal the latent patterns of a signal by decomposing it into low- and high-frequency bands and achieve multiresolution analysis by performing further decomposition over the low-frequency band. For our data, the wavelet transform could obtain the rough pattern of features for anatomical structures from the low-frequency band, and fine details from the high-frequency bands. Rather than transforming our input data directly, we propose to use this frequency information to improve calibration across views (see calibration details below). The stationary wavelet transform (SWT) [17] is an improved version of the discrete wavelet transform (DWT) and has different de-composition strategies. As shown in Fig. 3, SWT upsamples its filters by a factor of $2^{(i-1)}$ at the i-th level, which allows the output to retain the original resolution and avoid aliasing artifacts due to down-sampling of DWT [11]. In addition, the denoising property of SWT also helps to remove each view's irrelevant information. However, the output of SWT could be "redundant" due to the absence of downsampling. To mitigate this, we add a linear projection layer to compress the decomposed features from SWT. As each level of SWT generates a low- and a high-frequency subband that is of the same size as Λ, the output of a 3-level SWT is $S \in \mathbb{R}^{(6 \times c) \times n}$. Note that we concatenate each level's output according to the channel. Then, a linear projection is performed upon channel dimension to compress S to $S^* \in \mathbb{R}^{c \times n}$. This facilitates the fusion of low- and high-frequency features from different scales.

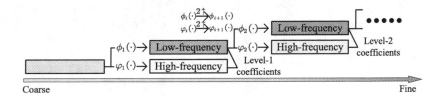

Fig. 3. The process of SWT decomposition. Here the low-pass and high-pass filters $\phi(\cdot)$ and $\varphi(\cdot)$ are upsampled after each level of decomposition.

Calibration. Since each anatomical view could provide unique information and could be im-portant for prediction, instead of directly applying the decomposed information from SWT for each view as the new input, we apply it to calibrate each view's original information. Motivated by [23] that produced attention from the discrete cosine transform, we apply Sigmoid to process the frequency information from S^*. The resultant score AS is then applied to calibrate each view-specific information V^i, as in the below equation.

$$AS = Sigmoid(S^*), \quad V^i = V^i \times AS \times \gamma + \beta \tag{1}$$

where γ and β are two learnable parameters incorporated to ease optimization [29]. The produced AS also provides interpretability that explains which anatomical region (the feature dimension of V^i) has higher contribution to the calibration.

3   Experiments

3.1   Experimental Settings

We evaluate the proposed model using Pytorch 1.12.1 and Nvidia RTX3090 card. All tests are based on 10-fold cross-validation. Before model training, L-2 normalization is applied to project the features into the same scale. The optimizer we choose is AdamW, with a weight decay of 0.05 and a learning rate of 0.01. To facilitate the convergence, we employ the cosine annealing learning rate scheduler. The loss is mean-squared loss. For SWT, we choose bior2.4 as the mother wavelet.

3.2   State-of-the-Art and Baseline Comparison

We choose the single-view backbone (Single-view) as a baseline and train it by con-catenating all views and treating the concatenated features as one view. This is a common comparison in multi-view learning [26]. Furthermore, we remove the cross-view calibrator from EMV-Net (w/o cross-view calibrator) to assess its effectiveness. Additionally, we add three ML-based models that were proposed for brain-based prediction, including SFCN [21], MLP [3], and ridge regression [14]. We follow the original designs and parameters suggested by the authors. Here we include the WM dataset and the GM dataset and test models' performances on age (demographic), strength (motor), and picture vocabulary (cognition) prediction. The results are shown in Table 1. Note that the metric for age prediction is the mean absolute error (MAE) and the metric for the other two is the correlation coefficient, which are popular metrics for the three tasks [3,21,28]. We further perform the repeated-measure ANOVA test and then the paired t-test to demonstrate the significance of the proposed EMV-Net when compared with other baselines and methods, as in Table 1.

3.3   Ablation Study

As presented in Table 2, we further show that the self-attention with 1.5-Entmax could outperform other activations (Softmax and Sparsemax). In addition, we compare different designs of the cross-view calibrator on whether SWT is applied (No SWT) or other levels of SWT is applied (level 1 and 2). All comparisons are based on the WM dataset. Based on the results, we test the statistical significance of our design. For the two groups in Table 2, repeated-measure ANOVA tests indicate significant differences. Furthermore, we do paired t-tests and demonstrate that our design significantly outperforms alternatives.

Table 1. Compare EMV-Net with and without the cross-view calibrator (in light gray). Also, compare the single-view baseline (in light gray) and three state-of-the-art methods in brain-based prediction (in dark gray). Note that the metric for age prediction is MAE, and the metric for strength and picture vocabulary prediction is the Spearman correlation coefficient. The paired t-test results for comparative implementations against the proposed one (EMV-Net) are presented by asterisks. * indicates that $p < 0.05$, and ** indicates that $p < 0.001$.

	EMV-Net			w/o Cross-view Calibrator			Single-view		
	Age	Strength	PicVocab	Age	Strength	PicVocab	Age	Strength	PicVocab
WM	2.51	0.69	0.38	2.61**	0.66*	0.36*	2.87**	0.60**	0.35*
GM	2.80	0.56	0.39	2.86*	0.55	0.38	3.03**	0.53*	0.36*
	SFCN			MLP			Ridge Regression		
	Age	Strength	PicVocab	Age	Strength	PicVocab	Age	Strength	PicVocab
WM	2.75**	0.56**	0.35*	2.98**	0.53**	0.19**	2.90**	0.61**	0.31**
GM	2.94**	0.54	0.37	3.06**	0.30**	0.25**	2.83*	0.52*	0.28**

Table 2. Ablation studies over the (1) activation functions for self-attention (compared to using 1.5-Entmax). (2) SWT for the cross-view calibrator. The paired t-test results for comparative implementations against the proposed one are presented by asterisks. * indicates that $p < 0.05$, and ** indicates that $p < 0.001$.

	Proposed	(1)		(2)		
		Sparsemax	Softmax	No SWT	SWT Level 1	SWT Level 2
Age	2.51	2.54*	2.73**	2.65*	2.58*	2.56*
Strength	0.69	0.68	0.65*	0.60*	0.68	0.69
PicVocab	0.38	0.35*	0.28**	0.30**	0.36	0.36

3.4   Interpretability Analysis

Compare Different Anatomical Views. We further analyze each anatomical view's importance for the three prediction tasks. We calculate each anatomical view's attention weights that produced from the view-specific feature highlighting phase of three cross-view calibrators based on the WM and GM feature datasets. Each view contains 1065 weights for the 1065 subjects. We average the weights and present the results as box plots.

Figure 4 shows that the model assigns different weights to various views for the three tasks, indicating a focus on distinct dMRI measures for prediction. For age prediction, FA1 variance and FA2 mean are the most important, while Trace2 median is the least important. The strength prediction has a more balanced focus on the 22 views compared to the other two tasks, with less variance in weights across these views. Picture vocabulary prediction relies heavily on FA1 min, which is less important for the other two tasks. Furthermore, the number of streamlines is the least important for picture vocabulary prediction but more important for the other two tasks.

The results on prediction based on the GM data is shown in Fig. 5. From the figure, surface thickness is more important for age and picture vocabulary

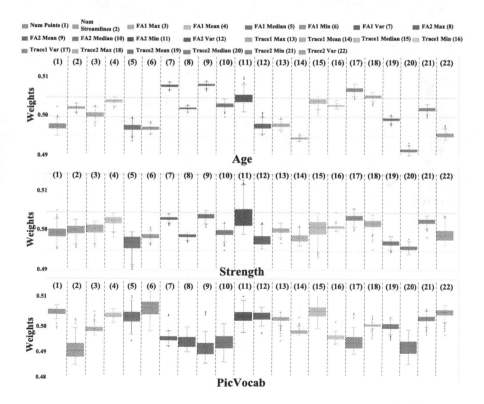

Fig. 4. Normalized weights for 22 anatomical views based on the WM dataset.

prediction. For strength prediction, both anatomical views are almost equal in importance.

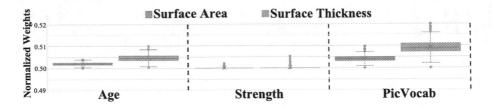

Fig. 5. The normalized weights for the 2 anatomical views based on the GM dataset.

4 Conclusion

In this study, we investigated a specific type of multi-view data that commonly oc-curs in segmented or parcellated MRI, where each anatomical structure is described with multiple features. We proposed to call this type of data anatomical-multi-view data. To efficiently extract useful features from multiple views while preserving interpretability, we have proposed an explainable multi-view network to learn both view-specific and across-view information. The model performed well on three different prediction tasks using WM and GM datasets. We further presented interpretable analyses of the importance of different anatomical views over the three tasks. We found that different anatomical measures of the brain white and grey matter have varied importance for predicting different human phenotypes. Future work may ex-tend this line of investigation to assess the effectiveness of EMV-Net on additional modalities of anatomical-multi-view-data, such as information from functional MRI. Overall, this investigation suggests that the exploration of methods designed specifically for anatomical-multi-view data holds potential for the study of the brain using machine learning.

5 Declaration of Conflict

There are no financial or non-financial conflict of interest.

References

1. Basser, P.J., Pajevic, S., Pierpaoli, C., Duda, J., Aldroubi, A.: In vivo fiber tractography using DT-MRI data. Magn. Reson. Med. **44**(4), 625–632 (2000)
2. Fadnavis, S., Polosecki, P., Garyfallidis, E., Castro, E., Cecchi, G.: MVD-Fuse: detection of white matter degeneration via multi-view learning of diffusion microstructure. bioRxiv 2021–04 (2021)

3. Feng, G., et al.: Methodological evaluation of individual cognitive prediction based on the brain white matter structural connectome. Hum. Brain Mapp. **43**(12), 3775–3791 (2022)
4. Fischl, B.: Freesurfer. Neuroimage **62**(2), 774–781 (2012)
5. Guo, J., et al.: CMT: convolutional neural networks meet vision transformers. In: Proceedings of the IEEE/CVF Conference on Computer Vision and Pattern Recognition, pp. 12175–12185 (2022)
6. Hackmack, K., Paul, F., Weygandt, M., Allefeld, C., Haynes, J.D., Initiative, A.D.N., et al.: Multi-scale classification of disease using structural MRI and wavelet transform. Neuroimage **62**(1), 48–58 (2012)
7. Hong, D., Huang, C., Yang, C., Li, J., Qian, Y., Cai, C.: FFA-DMRI: a network based on feature fusion and attention mechanism for brain MRI denoising. Front. Neurosci. **14**, 577937 (2020)
8. Hu, J., Shen, L., Sun, G.: Squeeze-and-excitation networks. In: Proceedings of the IEEE Conference on Computer Vision and Pattern Recognition, pp. 7132–7141 (2018)
9. Li, Y., Wu, F.X., Ngom, A.: A review on machine learning principles for multi-view biological data integration. Brief. Bioinform. **19**(2), 325–340 (2018)
10. Liem, F., et al.: Predicting brain-age from multimodal imaging data captures cognitive impairment. Neuroimage **148**, 179–188 (2017)
11. Mahabadi, A.A., Eshghi, M.: Speech enhancement using affine projection algorithm in subband. In: 2009 International Conference on Multimedia Computing and Systems, pp. 222–226. IEEE (2009)
12. Malcolm, J.G., Shenton, M.E., Rathi, Y.: Filtered multitensor tractography. IEEE Trans. Med. Imaging **29**(9), 1664–1675 (2010)
13. Mallat, S.: A Wavelet Tour of Signal Processing. Elsevier (1999)
14. Massett, R.J., et al.: Regional neuroanatomic effects on brain age inferred using magnetic resonance imaging and ridge regression. J. Gerontol.: Ser. A **78**(6), 872–881 (2023)
15. Mathew, A.R., Anto, P.B.: Tumor detection and classification of MRI brain image using wavelet transform and SVM. In: 2017 International Conference on Signal Processing and Communication (ICSPC), pp. 75–78. IEEE (2017)
16. Mota, B., Herculano-Houzel, S.: Cortical folding scales universally with surface area and thickness, not number of neurons. Science **349**(6243), 74–77 (2015)
17. Nason, G.P., Silverman, B.W.: The stationary wavelet transform and some statistical applications. In: Antoniadis, A., Oppenheim, G. (eds.) Wavelets and Statistics. LNCS, vol. 103, pp. 281–299. Springer, Cham (1995)
18. Niu, X., Zhang, F., Kounios, J., Liang, H.: Improved prediction of brain age using multimodal neuroimaging data. Hum. Brain Mapp. **41**(6), 1626–1643 (2020)
19. O'Donnell, L.J., Wells, W.M., Golby, A.J., Westin, C.-F.: Unbiased groupwise registration of white matter tractography. In: Ayache, N., Delingette, H., Golland, P., Mori, K. (eds.) MICCAI 2012. LNCS, vol. 7512, pp. 123–130. Springer, Heidelberg (2012). https://doi.org/10.1007/978-3-642-33454-2_16
20. O'Donnell, L.J., Westin, C.F.: An introduction to diffusion tensor image analysis. Neurosurg. Clin. **22**(2), 185–196 (2011)
21. Peng, H., Gong, W., Beckmann, C.F., Vedaldi, A., Smith, S.M.: Accurate brain age prediction with lightweight deep neural networks. Med. Image Anal. **68**, 101871 (2021)
22. Peters, B., Niculae, V., Martins, A.F.: Sparse sequence-to-sequence models. arXiv preprint arXiv:1905.05702 (2019)

23. Qin, Z., Zhang, P., Wu, F., Li, X.: FcaNet: frequency channel attention networks. In: Proceedings of the IEEE/CVF International Conference on Computer Vision, pp. 783–792 (2021)

24. Schmidt, A., Sharghi, A., Haugerud, H., Oh, D., Mohareri, O.: Multi-view surgical video action detection via mixed global view attention. In: de Bruijne, M., et al. (eds.) MICCAI 2021. LNCS, vol. 12904, pp. 626–635. Springer, Cham (2021). https://doi.org/10.1007/978-3-030-87202-1_60

25. Sun, S.: A survey of multi-view machine learning. Neural Comput. Appl. **23**, 2031–2038 (2013)

26. van Tulder, G., Tong, Y., Marchiori, E.: Multi-view analysis of unregistered medical images using cross-view transformers. In: de Bruijne, M., et al. (eds.) MICCAI 2021. LNCS, vol. 12903, pp. 104–113. Springer, Cham (2021). https://doi.org/10.1007/978-3-030-87199-4_10

27. Van der Velden, B.H., Kuijf, H.J., Gilhuijs, K.G., Viergever, M.A.: Explainable artificial intelligence (XAI) in deep learning-based medical image analysis. Med. Image Anal. **79**, 102470 (2022)

28. Weber, K.A., et al.: Confounds in neuroimaging: a clear case of sex as a confound in brain-based prediction. Front. Neurol. **13**, 960760 (2022)

29. Woo, S., et al.: ConvNext v2: co-designing and scaling convnets with masked autoencoders. In: Proceedings of the IEEE/CVF Conference on Computer Vision and Pattern Recognition, pp. 16133–16142 (2023)

30. WU-Minn, H.: 1200 subjects data release reference manual, **565** (2017). https://www.humanconnectome.org/

31. Yan, X., Hu, S., Mao, Y., Ye, Y., Yu, H.: Deep multi-view learning methods: a review. Neurocomputing **448**, 106–129 (2021)

32. Yang, Y., Wang, H.: Multi-view clustering: a survey. Big data mining and analytics **1**(2), 83–107 (2018)

33. Yuan, Y., Xun, G., Jia, K., Zhang, A.: A multi-view deep learning framework for EEG seizure detection. IEEE J. Biomed. Health Inform. **23**(1), 83–94 (2018)

34. Zekelman, L.R., et al.: White matter association tracts underlying language and theory of mind: an investigation of 809 brains from the human connectome project. Neuroimage **246**, 118739 (2022)

35. Zhang, F., et al.: Quantitative mapping of the brain's structural connectivity using diffusion MRI tractography: a review. Neuroimage **249**, 118870 (2022)

36. Zhang, F., et al.: SlicerDMRI: diffusion MRI and tractography research software for brain cancer surgery planning and visualization. JCO Clin. Cancer Inf. **4**, 299–309 (2020)

37. Zhang, F., et al.: Whole brain white matter connectivity analysis using machine learning: an application to autism. Neuroimage **172**, 826–837 (2018)

38. Zhang, F., et al.: An anatomically curated fiber clustering white matter atlas for consistent white matter tract parcellation across the lifespan. Neuroimage **179**, 429–447 (2018)

ReTrace: Topological Evaluation of White Matter Tractography Algorithms Using Reeb Graphs

S. Shailja[1(✉)], Jefferson W. Chen[2], Scott T. Grafton[1], and B.S. Manjunath[1]

[1] University of California, Santa Barbara, CA 93117, USA
shailja@ucsb.edu
[2] Irvine Medical Center, University of California, Santa Barbara, CA 92868, USA

Abstract. We present ReTrace, a novel graph matching-based topological evaluation and validation method for tractography algorithms. ReTrace uses a Reeb graph whose nodes and edges capture the topology of white matter fiber bundles. We evaluate the performance of 96 algorithms from the ISMRM Tractography Challenge and the standard algorithms implemented in DSI Studio for the population-averaged Human Connectome Project (HCP) dataset. The existing evaluation metrics such as the f-score, bundle overlap, and bundle overreach fail to account for fiber continuity resulting in high scores even for broken fibers, branching artifacts, and mis-tracked fiber crossing. In contrast, we show that ReTrace effectively penalizes the incorrect tracking of fibers within bundles while concurrently pinpointing positions with significant deviation from the ground truth. Based on our analysis of ISMRM challenge data, we find that no single algorithm consistently outperforms others across all known white matter fiber bundles, highlighting the limitations of the current tractography methods. We also observe that deterministic tractography algorithms perform better in tracking the fundamental properties of fiber bundles, specifically merging and splitting, compared to probabilistic tractography. We compare different algorithmic approaches for a given bundle to highlight the specific characteristics that contribute to successful tracking, thus providing topological insights into the development of advanced tractography algorithms.

Keywords: tractography · dMRI · Reeb graphs · topology · evaluation · graph matching

1 Introduction

Tractography is a technique utilized in neuroimaging to reconstruct white matter fiber pathways from diffusion magnetic resonance images (dMRIs). The reconstructed fiber bundles provide valuable insights into the connectivity between different regions of the brain. They play a crucial role in understanding

Supported by NSF award: SSI # 1664172.

neuroanatomy and studying various brain disorders. For instance, tractography has been used to reveal brain abnormalities across a range of conditions [1, 4, 9], including multiple sclerosis, cognitive disorders, Parkinson's disease, brain trauma, tumors, and psychiatric conditions. To ensure accurate interpretation of the obtained tractography results, it is vital to evaluate the performance of tractography methods on these neuroanatomical bundles and select appropriate metrics for assessment. Evaluating the performance of tractography methods is a complex task due to the intricate nature of white matter pathways and the challenges associated with capturing their neuroanatomical topology [5, 8].

Empirical studies have examined the effectiveness of many tractography methods in investigating a range of neurodegenerative diseases like Alzheimer's [19], as well as in measuring patient outcomes following the utilization of tractography for tumor resection [18]. However, the comparison between tractography algorithms remains mainly qualitative for the most part. These studies do not provide a comprehensive evaluation of the tractography algorithms themselves, as they lack a ground truth reference for the reconstructed bundles. This limitation hinders the ability to quantitatively compare standard tracking algorithms and obtain meaningful feedback about their performance.

To quantify tractography, the FiberCup phantom dataset [3] is commonly used. The International Society for Magnetic Resonance in Medicine (ISMRM) organized a tractography challenge [10] on revised FiberCup dataset to establish a ground truth and provide score using the Tractometer [2]. Tractometer provides global connectivity metrics and facilitates extensive assessment of tracking outputs, fiber bundle detection accuracy, and incomplete fiber quantification. As bundle analysis is crucial in neurological studies, the tractograms were divided into 25 major bundles in the ISMRM dataset. To assess the performance of tractography method on bundles, bundle coverage metrics were proposed [10]. These metrics transform the fibers into voxel images, which results in the loss of fiber point-correspondence. They fail to account for many reconstruction errors, thereby yielding inflated and potentially misleading scores. For example, the topological complexity arising due to the geometrical structure and branching within valid bundles is often neglected in voxel-based metrics. In other words, the existing tractography assessment metrics do not answer "how" the fibers are connected but only analyze the connection percentages. Hence, there is a critical need for methods that can quantify the anatomical validity of fiber branching, given the complex fiber topology [17].

Topology pertains to the organization of white matter fibers into structural networks of brain. It considers the fibers' origination, termination, and branching, as well as their relationship to different brain regions. Neurological disorders can induce topological changes in white matter fibers, such as physical disruptions in cases like stroke, or spatial distortions as seen in brain tumors. To effectively assess the topology of the fiber reconstruction in bundle tracts, we propose ReTrace. It is a novel graph matching algorithm that is based on the construction of a Reeb graph [14–16]. This graph matching algorithm enables a comprehensive quantitative analysis of topological connectivity patterns by considering both global and local network features. Further, for each quantitative

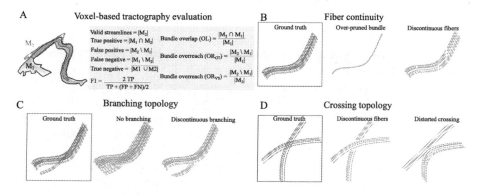

Fig. 1. Traditional bundle coverage metrics and their limitations in tractography evaluation. (A) Mask M_1 illustrates the ground truth bundle (black outline), while Mask M_2 reveals valid fibers (red outline) produced by a specific tractography method. Bundle coverage evaluation uses these voxel masks. (B)–(D) display scenarios where the traditional metric falls short. Gray fibers represent the ground truth, while orange fibers depict the reconstructed fibers. (B) highlights potential errors in fiber continuity. (C) demonstrates potential inaccuracies in the branching topology of the tracked fibers, and (D) exposes potential misinterpretations in crossing topology, which despite yielding high scores in bundle coverage, might be anatomically implausible. (Color figure online)

assessment provided by ReTrace, a graph visualization of the bundle in 3D is also associated. This visualization is crucial in analyzing the efficacy of the tractography. Finally, ReTrace can also be tuned to explore the output of a given tractogram in different resolutions. The implementation code and interactive notebooks for utilizing ReTrace are publicly available on GitHub[1].

2   Tractography Evaluation

Quantitative metrics such as global connectivity-dependent metrics in Tractometer [2] and connectivity matrix have been used to evaluate the fiber reconstruction [13]. They give information about the high-level connection of brain regions. However for the assessment of a given bundle tract, they offer a snapshot of valid bundle coverage with their ground truth counterparts and do not account for local topological features. We describe the commonly used metrics in Fig. 1A. While the existing metrics are valuable in assessing volumetric bundle coverage they fall short in capturing fiber continuity, branching, and crossing as highlighted in Fig. 1B, C, and D respectively in the bundle. Some of these limitations are:

1. Voxel-wise agreement between the algorithm's output and the ground truth is the main emphasis in computing the existing metrics. This overlooks the intricate tractography errors, such as, spurious fibers and incorrect connectivity patterns. Specifically, these metrics often neglect broken fibers as they could still cover the volume despite their discontinuity.

---

[1] https://github.com/s-shailja/ReTrace.

2. Spatial localization is not considered to evaluate the errors between the reconstructed bundles and the ground truth. Similarly, tractography errors cannot be attributed to specific inaccuracies in tracked fibers within the spatial context to guide the design of better algorithms.
3. Branching in the reconstructed pathways is largely inconsequential in the computation of voxel-based metrics. Typically, fiber bundles originate from functional regions, merge into larger pathways for efficiency, and branch out as they approach their respective terminal functional areas. Neuroanatomical studies [5, 15] have pointed out the importance of branching in tractography. But, the conventional metrics (in Fig. 1A) may fail to detect errors related to this branching topology. As such, obscured connections within dense bundles could still achieve high scores despite inaccuracies.
4. Complex fiber orientations such as fiber crossing are not directly captured by the existing tractography evaluation metrics.

Our proposed method addresses these limitations for evaluating tractography methods. It incorporates higher-level tract analysis that considers the topological branching patterns and pathways. Additionally, our method is sensitive to spatial localization, taking into account the precise anatomical locations of inaccuracies within the reconstructed bundles. By addressing the limitations discussed above, our method provides a tunable and robust evaluation of tractography algorithms in terms of accuracy and anatomical fidelity.

3   ReTrace: Quantifying Topological Analysis

ReTrace is an end-to-end tractography evaluation pipeline that starts by processing the given white matter fiber bundles to generate a graph model. Then, it evaluates the quality of fiber reconstruction by comparing two graphs taking into account many topological factors. The pipeline is illustrated in Fig. 2. It is based on the construction of a Reeb graph that provides a topological signature of the fibers. This graph representation makes the tractography evaluation amenable to graph and network theory methods.

3.1   Reeb Graphs Characterize the Topology of Anatomical Bundles

A Reeb graph representation of white matter fibers has been recently proposed [14–16]. It provides a concise depiction of trajectory branching structures by constructing undirected weighted graphs. The given bundle tracts (that is, a group of streamlines representing a major bundle) is represented as a graph. The vertices of the graph are critical points where the fibers appear, disappear, merge, or split. Edges of the graph connect these critical points as groups of subtrajectories of the given bundle. The edge weight is the proportion of fibers that participate in the edge. Three key parameters ($\epsilon, \alpha,$ and δ) capture the geometry and topology of fibers. ϵ denotes the distance between a pair of fibers in a bundle, controlling its sparsity. Smaller ϵ values result in denser subtrajectory groups, while larger values allow for sparser groups. α represents the spatial length of the

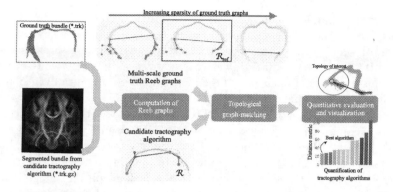

Fig. 2. ReTrace evaluates tractography methods both qualitatively and quantitatively by comparing two Reeb graphs, $\mathcal{R}$ and $\mathcal{R}_{\text{ref}}$. By setting appropriate parameters, Reeb graphs of different resolutions can be computed from the ground truth data. Among the graphs of different sparsity, one is chosen based on the level at which tractography evaluation is desired. Similarly, for the candidate tractography algorithm under evaluation, a Reeb graph is computed from the bundle using the same Reeb graph parameters as the ground truth. The evaluation metric reflects the distance between the two graphs: a higher metric value indicates larger discrepancies, and the candidate algorithm with the lowest metric is assigned the first rank. For qualitative analysis, one can visually examine the Reeb graphs, locating nodes contributing to the distance. (Color figure online)

bundle, introducing persistence and influencing the extent of bundling. δ determines the bundle thickness to shape the model's robustness and granularity. By adjusting these parameters, we can explore the anatomical fiber structure at different scales of sparsity. For example, the *branching* structure is finely encoded when the inter-trajectory distance, ϵ is less. The interruption tolerance can be enhanced by increasing α. Finally, to ignore spurious fibers and bundles δ can be increased in the Reeb graph. In this paper, we analyze the reconstruction of anatomical bundles in different spatial resolutions, demonstrating the potential of graph-based tractography validation and evaluation.

Note: Throughout the paper, we overlay the Reeb graphs on raw fibers (in orange and ground truth fibers in grey). In the graphs, the nodes are illustrated in red, while edges are shown in black.

3.2 Topological Graph Matching

For any given node $v \in V$, we calculate two sets of features—spatial position-based features, denoted as "pos", and local network-level features, denoted as "net". Each node of the Reeb graph is linked to its 3D spatial location in the brain, so "pos" corresponds to the 3D coordinate yielding a 3-dimensional feature. On the other hand, the network features are computed using centrality metrics at the node level: degree centrality, closeness centrality, betweenness centrality, and eigenvector centrality [6], producing a 4-dimensional feature.

Our graph matching computation algorithm is adapted from Siminet [11], enhanced to accommodate additional parameters for robustness against noise in tractography and to incorporate network metrics with spatial position as node features. It calculates an edit distance denoted by d_{edit} between two Reeb graphs, quantifying the cost of operations needed to transform one graph into another, as outlined in Algorithm 1 in the Appendix 2.

The algorithm begins by iterating through the nodes of the comparison graph $\mathcal{R}$ and matching each node with its closest spatial counterpart within a search radius. The algorithm determines node correspondences by comparing the spatial distances ("pos") between nodes and their counterparts against a threshold ϵ (inter-fiber distance used in constructing the Reeb graph). The algorithm also accounts for insertions and deletions by penalizing them with a score γ proportional to the Euclidean distance between the centroids of node locations for $\mathcal{R}$ and $\mathcal{R}_{ref}$. The proposed metric d_{edit} is computed considering the Euclidean distance in node attributes: spatial positions and centrality metrics to compare graph-based representations of white matter bundles, ensuring that layout similarities correspond to similarities in brain regions.

Limitations: Note that the metric is not normalized, that is, it does not have the standard 0–1 limits. We opt to keep the distances between graphs unbounded to account for relative variations (in some cases, the distance could be very large). The ultimate aim of a new tractography design would be to minimize this distance. A potential limitation, or rather a feature, is that the user needs to select the resolution parameters based on the desired comparison of the bundle tracts. While this might appear a manual step, the preset parameters generally serve well as a great starting point for all the major bundle tracts. Also, the performance of all tractography evaluation metrics, both existing and our proposed metric, depends on the quality of bundle segmentation. Hence, further research on bundle segmentation could enhance metric performance.

4   Results

4.1   The ISMRM 2015 Tractography Challenge

The challenge presented participants with a clinical-style dataset: a 2mm isotropic diffusion acquisition with 32 gradient directions and a b-value of $1000\,\mathrm{s/mm^2}$. The task was to reconstruct fiber pathways using a realistically simulated replication of a whole-brain diffusion-weighted MR image. The challenge resulted in 96 tractogram submissions, available publicly for download[2]. These submissions collectively represent a broad range of tractography pipelines, encompassing varied pre-processing, tractography, and post-processing algorithms. This provides a diverse platform for quantitative and qualitative analysis of tractography methods. We assign each algorithm a unique algorithm number, ranging from 1–96 (see Appendix 2 for the mapping of these ids to the original submission numbers).

---

[2] https://zenodo.org/record/840086.

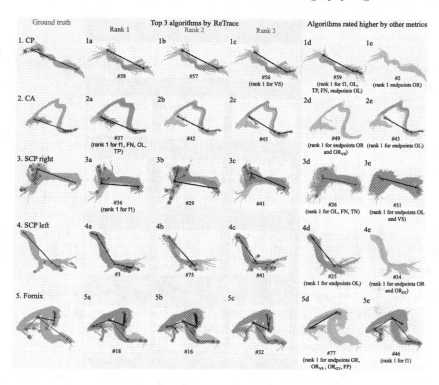

Fig. 3. Evaluation of tractography algorithms using ReTrace for bundles in the ISMRM data. The leftmost column displays the ISMRM ground truth fiber bundles with associated Reeb graphs overlaid on the bundles. For the other columns, grey streamlines denote the ground truth, overlaid by orange streamlines representing the candidate tractography algorithm's results. On top of these, the Reeb graphs are overlaid with red nodes and black edges. The bundles illustrated in the figure are labeled from 1 to 5: CP, CA, SCP$_{right}$, SCP$_{left}$, and Fornix respectively. For each bundle (in a row), the next three columns (a-c) highlight the algorithms that achieved the highest rank according to our proposed metric. Notice that the algorithms that are ranked high on existing metrics (displayed in columns d-e), do not show the best tracked bundle. The reconstructions that do not capture the branches near the end of the bundles are ranked highly by other metrics. Similarly, fiber continuity and distorted fibers in the reconstructed output are given high ranks with existing metrics. (Color figure online)

For tractography evaluation, we compute the valid bundles for each submission using the ROI-based bundle segmentation system proposed by the challenge organizers [12]. An expert segmented ground truth bundle segmentation provided by the challenge allowed us to assess the submitted tractography methods based on traditionally computed metrics, establishing a baseline for comparison. The extracted bundles were then processed using the ReTrace pipeline. We can fine-tune the computation of Reeb graphs using the robustness parameters (ϵ, α, and δ), adjusting the granularity of the desired analysis. In this study, these are set to $\epsilon = 2.5$, $\alpha = 5$, and $\delta = 5$. The results for various bundles using our proposed

Table 1. Comparison of d_{edit} with existing bundle coverage metrics. The columns labeled as "#" represent the rank 1 algorithm number, while "val" denotes the respective metric values for each metric. Top-ranking methods according to d_{edit} differ from those determined by traditional metrics for most bundles. We demonstrate that the reconstruction ranked the best using d_{edit} effectively captures branching topology that is overlooked in evaluations using existing methods.

Bundle	FN		FP		OL		OR$_{\text{GT}}$		OR$_{\text{VS}}$		TP		VS		Endpoints OL		Endpoints OR		f1		d_{edit}	
	#	val	#	val	#	val	#	val	#	val	#	val	#	val	#	val	#	val	#	val	#	val
CP	59	1134	75	38	59	0.29	75	0.02	32	0.42	59	474	56	34	59	14	42	1	59	0.33	58	40.1
CA	37	603	60	72	37	0.59	60	0.05	49	0.43	37	876	38	75	43	22	49	2	37	0.52	37	7.3
SCP right	26	471	44	6	26	0.88	44	0	44	0.08	26	3467	57	1768	57	242	44	1	36	0.57	36	10.6
SCP left	26	440	34	12	26	0.89	34	0	77	0.05	26	3638	57	2245	25	352	34	0	16	0.63	3	10.9
Fomix	26	2698	77	40	26	0.75	77	0	77	0.07	26	7971	25	2595	25	698	77	3	46	0.57	18	8.7

metric compared with the traditional bundle coverage metrics are illustrated in Fig. 3. When the candidate tractography has less than 5 fibers, Reeb graphs are not constructed (automatically with the Reeb graph parameter $\delta = 5$) and the algorithm is rejected without rank assignment. For a more dense analysis δ may be set to 0. The comparison of existing bundle coverage metrics against our proposed d_{edit} metric is presented in Table 1.

All submissions faced difficulties reconstructing the smaller bundles, such as the anterior (CA) and posterior commissures (CP), which possess a cross-sectional diameter of no more than 2 mm. Due to the minimal branching owing to fewer recovered fibers, ReTrace performs similar to the existing methods on these bundles. However, as the number of fibers increases for bundles like SCP$_{\text{right}}$, SCP$_{\text{left}}$, and Fornix (as shown in Fig. 3), the importance of accurately branching towards the end of the fibers became apparent. This leads to different algorithms being top-rank with our method as compared with existing metrics. As an example, while algorithm 36 and 29 achieve high ranks for the ReTrace evaluation in SCP$_{\text{right}}$, they do not perform as well according to the existing metrics. It is notable that simply increasing fiber count to cover the bundle volume without appropriate branching does not lead to a high ReTrace ranking. This is evident in algorithms 26 and 31 (Fig. 3d and 3e) as they are ranked high according to OL, FN, TN, endpoints OL, and VS metrics, but do not contribute significantly to the ReTrace ranking. The evaluation of all other bundles in the ISMRM dataset using d_{edit} along with their interactive visualizations are available on our GitHub repository. Beyond the topological distance that we calculate here, various other graph attributes such as variations in the number of nodes, edges, average degree, and diameter can also be computed and are available in our repository[3]. They collectively capture different facets of the graphs, and as a consequence, highlight various properties of the tractography reconstruction.

Different tractography methods were employed by different teams. The most important preprocessing, postprocessing, and tractography steps among top-performing algorithms are shown in Fig. 4. For example, CP proved difficult to

---

[3] https://github.com/s-shailja/ReTrace.

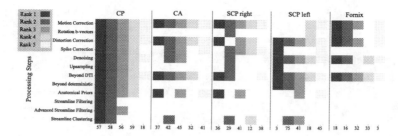

Fig. 4. Correlation between processing steps and the successful reconstruction of the major bundles as assessed by ReTrace. Different colors represent the ranking of methods according to the ReTrace pipeline for five different bundles (rank 1 (dark blue) to rank 5 (yellow)). Y-axis indicate various steps of tractography: Preprocessing (from motion correction to upsampling); Tractography (diffusion modeling beyond DTI such as constrained spherical deconvolution, and tractography methods beyond deterministic approaches such as probabilistic tractography); Postprocessing (incorporation of anatomical priors to streamline clustering). The importance of each step is evident with how prevalent it is in the top ranked algorithms for each bundle. (Color figure online)

reconstruct for all algorithms, necessitating the use of extensive denoising and correction methods, as well as probabilistic methods for fiber tracking. Conversely, for other bundles, minimal preprocessing was required and deterministic fiber tracking was sufficient for accurate branching. The insights on the steps of the algorithms offer a topological perspective on designing advanced tractography algorithms and pipelines for future studies. We observe that none of the algorithms perform consistently well for all the bundles. So, an additional advantage of this exploration is that new tractography algorithms may be designed, tailored to specific bundles or neuroanatomical properties of interest.

The results in this paper are in contrast with the ranking of tractography algorithms discussed in the existing literature [10,12] but also confirm the findings of the importance of various steps in tractography [13]. The runtime of the pipeline depends on the number of fibers and the desired resolution of the Reeb graph. After computing the graphs for a bundle, our evaluation method takes less than 120 s to obtain the results on an Intel Xeon CPU E5-2696 v4 @ 2.20 GHz.

4.2   Evaluating Tractography Algorithms in Fiber-Crossing Regions

Tractography's accuracy is influenced by the number of distinct fiber orientations per voxel, as fiber-crossing regions pose considerable challenges. A probabilistic white matter atlas highlights these areas with high fiber-crossing frequencies [17]. Conventional metrics may not fully capture an algorithm's ability to accurately resolve these complex regions, instead reflecting its capacity to generate copious or elongated fibers, regardless of their neurological plausibility or accuracy.

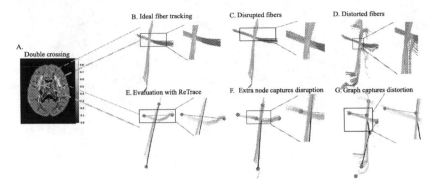

Fig. 5. Evaluation of a tractography algorithm for fiber crossing. (A) shows the center slice of the probabilistic double crossing atlas. A small ROI with high double crossing fiber probability (shown in green) is selected. (B)–(D) demonstrate the reconstruction of fibers using the streamline method implemented in DSI Studio. (B) shows successful tracking of fiber crossings with ideal parameters. (C) shows fibers that terminate abruptly near the crossing as a result of reconstruction with different parameters. (D) shows bending or distortion of fibers when encountering multiple diffusion orientations. (E)-(G) present the constructed Reeb graphs overlaid on the computed fibers for each case, respectively. An additional node is observed in the Reeb graph that captures the fiber crossing or bending. A thicker edge in (G) indicates bending of the fibers, whereas ideally, the fiber should only cross without bending. (Color figure online)

For example, a tractography algorithm might generate spurious fibers connecting distinct bundles in crossing fiber scenarios, falsely inflating the fiber count. This inflation could distort performance evaluations if they rely solely on fiber numbers. Endpoint-matching metrics, such as overlap or intersection between reconstructed and ground truth fibers, could also be misleading in the presence of crossing fibers: overlaps may not indicate accuracy if the algorithm incorrectly generates disrupted or distorted fibers due to crossing fibers. Spatial accuracy metrics, like the average or Hausdorff distance [7] between reconstructed and ground truth fibers, could also be skewed by crossing fibers. These regions may compromise the algorithm's capability to accurately resolve individual bundles, resulting in spatial inaccuracies or increased distances. The efficacy of ReTrace is not limited to synthetic data alone. To demonstrate its applicability to real datasets and to show how Retrace handles fiber crossing, we use the average HCP 1065 template, constructed from the diffusion MRI data of 1065 subjects from the Human Connectome Project (HCP)[4] as shown in Fig. 5. The experimental setting can be found in the Appendix 2.

---

[4] https://brain.labsolver.org/hcp_template.html.

5   Conclusion

This paper introduces ReTrace, an innovative evaluation method for tractography algorithms. This method addresses existing metrics' limitations by focusing on the topological accuracy of reconstructed pathways. We applied ReTrace to both synthetic and real-world datasets to demonstrate the branching fidelity and errors such as broken or bent fibers in tractography reconstruction. With the ISMRM dataset, we ranked 96 algorithms on different neuroanatomical bundles. The rankings proposed by our method are in contrast with the rankings using the conventional voxel-based tractography metrics. We discuss this difference in ranking and performance of different algorithms by highlighting the topological features such as branching, fiber continuity, localization, and crossing. We demonstrated the utility of our method on the HCP dataset where we show the importance of reconstructed fiber crossing and discuss the performance of standard tractography algorithms. Our approach is disease-agnostic, does not require brain registration to an atlas, and works across different acquisition protocols. It is important to note that bundle segmentation is a common bottleneck in any evaluation metric. Therefore, advancements in segmentation research could greatly enhance these evaluations. The results on tractography comparison presented here could be extended to be used as a cost function for data-driven machine learning methods, like generative adversarial networks. With feedback from neuroscientists, we hope that the results in this paper will pave the way forward in improving existing tractography methods.

Acknowledgement. The authors acknowledge Vikram Bhagavatula for his preliminary implementation of the edit distance. Data were provided by The ISMRM 2015 Tractography Challenge and by HCP, WU-Minn Consortium (Principal Investigators: David Van Essen and Kamil Ugurbil; 1U54MH091657) funded by the 16 NIH Institutes and Centers that support the NIH Blueprint for Neuroscience Research; and by the McDonnell Center for Systems Neuroscience at Washington University.

Mapping of algorithm IDs to the original submission IDs:

Submission ID	Algorithm ID	Submission ID	Algorithm ID
1_0	1	10_19	49
1_1	2	11_0	50
1_2	3	11_1	51
1_3	4	12_0	52
1_4	5	12_1	53
2_0	6	12_2	54
3_0	7	12_3	55
3_1	8	13_0	56
3_2	9	13_1	57
3_3	10	13_2	58
3_4	11	13_3	59
4_0	12	14_0	60
5_0	13	14_1	61
5_1	14	14_2	62
6_0	15	15_0	63
6_1	16	16_0	64
6_2	17	16_1	65
6_3	18	16_2	66
6_4	19	16_3	67
7_0	20	16_4	68
7_1	21	17_0	69
7_2	22	17_1	70
7_3	23	17_2	71
8_0	24	17_3	72
9_0	25	17_4	73
9_1	26	18_0	74
9_2	27	18_1	75
9_3	28	18_2	76
9_4	29	18_3	77
10_0	30	18_4	78
10_1	31	19_0	79
10_2	32	19_1	80
10_3	33	19_2	81
10_4	34	20_0	82
10_5	35	20_1	83
10_6	36	20_2	84
10_7	37	20_3	85
10_8	38	20_4	86
10_9	39	20_5	87
10_10	40	20_6	88
10_11	41	20_7	89
10_12	42	20_8	90
10_13	43	20_9	91
10_14	44	20_10	92
10_15	45	20_11	93
10_16	46	20_12	94
10_17	47	20_13	95
10_18	48	20_14	96

Algorithm 1 Topological Distance Computation

---

function ONEWAYDISTANCE($\mathcal{R}, \mathcal{R}_{ref}, \epsilon, \gamma$)
  I $\leftarrow$ Vref   ▷ *Nodes to be inserted*
  D $\leftarrow \emptyset$ ▷ *Nodes to be deleted*
  $d_{edit} \leftarrow 0$ ▷ *Score*
  $d_{pos} \leftarrow 0$ ▷ *Spatial score*
  $d_{net} \leftarrow 0$ ▷ *Network score*
  for $n_c \in V$ **do**
    c $\leftarrow$ closest node in V_{ref}
    if $d(n_c["pos"], n_r["pos"]) < 2\epsilon$ **then**
      Remove c from I   ▷ *Successful correspondence*
      if $d(n_c["pos"], n_r["pos"]) < \epsilon$ **then**
        continue   ▷ *Equivalence (nothing added)*
      $d_{pos} \leftarrow d_{pos} + d(n_c["pos"], n_r["pos"])$
                                   ▷ *Substitution*
      $d_{net} \leftarrow d_{net} + d(n_c["net"], n_r["net"])$
    else
      Add c to D
  $d_{pos} \leftarrow d_{pos} + |I| \cdot 2\epsilon(1 + \gamma)$ ▷ *Insertion*
  for $n_i \in I$ **do**
    $d_{net} \leftarrow d_{net} + d(n_i, 0)$
  $d_{pos} \leftarrow d_{pos} + |D| \cdot 2\epsilon(1 + \gamma)$ ▷ *Deletion*
  for $n_d \in D$ **do**
    $d_{net} \leftarrow d_{net} + d(n_d, 0)$
  $d_{edit} \leftarrow d_{pos}/|V_{ref}| + d_{net}$
  return d_{edit}
function DISTANCE($\mathcal{R}, \mathcal{R}_{ref}, \epsilon, \gamma$)
  $s_{cmp} \leftarrow$ OneWayDistance($R, R_{ref}, \epsilon, \gamma$)
  $s_{ref} \leftarrow$ OneWayDistance($R_{ref}, R, \epsilon, \gamma$)
  return $0.5 \cdot (s_{cmp} + s_{ref})$

HCP Dataset

ReTrace handles fiber crossings effectively, as demonstrated in Fig. 5. We use the 1 mm population-averaged FIB file in the ICBM152 space for fiber tracking in DSI Studio. We select a small region of interest (a 2 mm isotropic 3D region) from the probabilistic atlas in an area with a high probability (~ 0.9) of double-crossing fibers. We use the deterministic streamline tracking method implemented in DSI Studio[5] to compute the fibers with the parameters set (angular threshold, step size, min length, max length, terminate if seeds, iterations for topological pruning) to 35, 1, 70, 200, 1000, and 16, respectively. This allowed us to observe successful tracking without broken or distorted fibers. To mimic the spurious broken fibers that tractography methods may yield, we set the parameters to 35, 1, 0, 100, 1000, and 16. For observing the angular distortion where the fiber bends and follows a different path, we set the parameters as 90, 1, 70, 100, 1000, and 16. The resulting Reeb graphs clearly highlight how their nodes capture successful and unsuccessful tracking. Nodes formed near intersections indicate broken or bent fibers, as shown in Fig. 5. Whenever fibers travel together in a group, they form an edge in the graph. Any alteration within this group prompts a critical event, resulting in a node in the Reeb graph. Consequently, if a fiber breaks or diverges, the associated group changes, generating a node. This node, present at the merging point, contributes to a larger distance value. In the topological distance computation, d_{edit}, this node could be weighted more if the goal is to assess an algorithm's tracking ability in ambiguous fiber orientations. By providing a 3D location for our algorithm's attention, any discrepancy within that location can significantly affect the overall d_{edit} computation. The code is open source and can be tailored to specific needs.

---

[5] https://dsi-studio.labsolver.org/

References

1. Arienzo, D., et al.: Abnormal brain network organization in body dysmorphic disorder. Neuropsychopharmacology **38**(6), 1130–1139 (2013)
2. Côté, M.A., Girard, G., Boré, A., Garyfallidis, E., Houde, J.C., Descoteaux, M.: Tractometer: towards validation of tractography pipelines. Med. Image Anal. **17**(7), 844–857 (2013)
3. Fillard, P., et al.: Quantitative evaluation of 10 tractography algorithms on a realistic diffusion MR phantom. Neuroimage **56**(1), 220–234 (2011)
4. García-Gomar, M., et al.: Probabilistic tractography of the posterior subthalamic area in Parkinson's disease patients. J. Biomed. Sci. Eng. (2013)
5. Gillard, J., et al.: MR diffusion tensor imaging of white matter tract disruption in stroke at 3T. Br. J. Radiol. **74**(883), 642–647 (2001)
6. Hagberg, A., Swart, P., S Chult, D.: Exploring network structure, dynamics, and function using networkx. Technical report, Los Alamos National Lab. (LANL), Los Alamos, NM (United States) (2008)
7. Huttenlocher, D.P., Klanderman, G.A., Rucklidge, W.J.: Comparing images using the Hausdorff distance. IEEE Trans. Pattern Anal. Mach. Intell. **15**(9), 850–863 (1993)
8. Jones, D.K.: Challenges and limitations of quantifying brain connectivity in vivo with diffusion MRI. Imaging Med. **2**(3), 341 (2010)
9. Kao, P.Y., et al.: Improving patch-based convolutional neural networks for MRI brain tumor segmentation by leveraging location information. Front. Neurosci. **13**, 1449 (2020)
10. Maier-Hein, K.H.: The challenge of mapping the human connectome based on diffusion tractography. Nat. Commun. **8**(1), 1349 (2017)
11. Mheich, A., Hassan, M., Khalil, M., Gripon, V., Dufor, O., Wendling, F.: SimiNet: a novel method for quantifying brain network similarity. IEEE Trans. Pattern Anal. Mach. Intell. **40**(9), 2238–2249 (2017)
12. Renauld, E., Théberge, A., Petit, L., Houde, J.C., Descoteaux, M.: Validate your white matter tractography algorithms with a reappraised ISMRM 2015 Tractography Challenge scoring system. Sci. Rep. **13**, 2347 (2023)
13. Sarwar, T., Ramamohanarao, K., Zalesky, A.: Mapping connectomes with diffusion MRI: deterministic or probabilistic tractography? Magn. Reson. Med. **81**(2), 1368–1384 (2019)
14. Shailja, S., Bhagavatula, V., Cieslak, M., Vettel, J.M., Grafton, S.T., Manjunath, B.S.: ReeBundle: a method for topological modeling of white matter pathways using diffusion MRI. IEEE Trans. Med. Imaging (2023). https://doi.org/10.1109/TMI.2023.3306049
15. Shailja, S., Grafton, S.T., Manjunath, B.: A robust Reeb graph model of white matter fibers with application to Alzheimer's disease progression. bioRxiv, 2022-03 (2022)
16. Shailja, S., Zhang, A., Manjunath, B.S.: A computational geometry approach for modeling neuronal fiber pathways. In: de Bruijne, M., et al. (eds.) MICCAI 2021. LNCS, vol. 12908, pp. 175–185. Springer, Cham (2021). https://doi.org/10.1007/978-3-030-87237-3_17
17. Volz, L.J., Cieslak, M., Grafton, S.: A probabilistic atlas of fiber crossings for variability reduction of anisotropy measures. Brain Struct. Funct. **223**, 635–651 (2018)

18. Wu, J.S.: Clinical evaluation and follow-up outcome of diffusion tensor imaging-based functional neuronavigation: a prospective, controlled study in patients with gliomas involving pyramidal tracts. Neurosurgery **61**(5), 935–949 (2007)
19. Zhan, L., et al.: Comparison of nine tractography algorithms for detecting abnormal structural brain networks in Alzheimer's disease. Front. Aging Neurosci. **7**, 48 (2015)

Advanced Diffusion MRI Modeling Sheds Light on FLAIR White Matter Hyperintensities in an Aging Cohort

Kelly Chang[1][✉], Luke Burke[4], Nina LaPiana[2], Bradley Howlett[2], David Hunt[2], Margaret Dezelar[4], Jalal B. Andre[3], James Ralston[4], Ariel Rokem[1], and Christine Mac Donald[2]

[1] Department of Psychology, University of Washington, Seattle, USA
{kchang4,arokem}@uw.edu
[2] Department of Neurological Surgery, University of Washington, Seattle, USA
cmacd@uw.edu
[3] Department of Radiology, University of Washington, Seattle, USA
[4] Kaiser Permanente Washington Health Research Institute, Seattle, USA

Abstract. White matter hyperintensities (WMH) in fluid-attenuated inversion recovery (FLAIR) imaging are used as an indicator of clinical conditions ranging from multiple sclerosis to cerebrovascular disease. However, the biophysics underlying FLAIR WMH is only partially understood. In contrast, advanced diffusion MRI (dMRI) modeling, such as multi-shell and high angular resolution imaging, provide biophysically interpretable tissue properties but is more costly to measure. To study the relationships between dMRI tissue properties and FLAIR WMH in aging, we applied unsupervised learning to characterize the distribution of tissue properties within WMH regions of interest (ROI) in a sample from the Adult Changes in Thought (ACT) study. A principal components analysis (PCA) of the dMRI metrics within WMH ROIs revealed that the first two PCs explained 86.84% of the variance in the dataset. Using a tractometry approach, we focused on the FLAIR signal along the callosal pathways connecting the temporal lobes, demonstrating that posterior periventricular WMH are related to the loss of axonal tissue and intrusion of CSF into the white matter.

Keywords: diffusional kurtosis imaging · free-water diffusion tensor imaging · FLAIR · unsupervised learning · principal component analysis · tractometry

1 Introduction

Multi-shell, high angular resolution, diffusion MRI (dMRI) provides a sensitive measurement of brain white matter tissue properties. Nevertheless, in clinical

---

A. Rokem and C. Mac Donald—These authors contributed equally.

M. Karaman et al. (Eds.): CDMRI 2023, LNCS 14328, pp. 192–203, 2023.
https://doi.org/10.1007/978-3-031-47292-3_17

settings, this type of detailed measurement is prohibitively lengthy and complicated to collect, and so white matter tissue properties are often visually assessed
in a fluid-attenuated inversion recovery (FLAIR) sequence instead. However,
although FLAIR measurements are useful as diagnostic tools when read by
expert neuroradiologists, they lack the quantitative, biophysical interpretation
offered by dMRI models. The goal of the present work is to link signs of aging
that appear in FLAIR measurements of white matter with biophysical models
of dMRI data. To do so, we take advantage of a dataset collected as part of the
Adult Changes in Thought (ACT) study. The overall goal of the ACT study
is to conduct research to understand factors that contribute to Alzheimer's
Disease and related dementia. The ACT study leverages a repository of care-
fully collected and curated data resources including self-report, electronic health
records, biologic, and device data to deepen the understanding of the aging brain
in a well-characterized, community-based, longitudinal, prospective cohort study
[21].

Here, we focus on research-grade MRI measurements conducted with a subset
of participants from the ACT study. Each individual in this subset participated
in a scan session that included measurements of FLAIR and multi-shell high
angular resolution (HARDI) dMRI measurements. An automated method [24]
was used to segment FLAIR measurements and delineate white matter hyperintensities (WMH) regions of interest (ROI). DMRI measurements were used to
model diffusion with diffusional kurtosis imaging (DKI) [18] and its biophysical
extension in the White Matter Tract Integrity (WMTI) model [11]. In addition, the dMRI signal was modeled using a two-compartment model consisting
of tissue-restricted water and freely-diffusing water (FWDTI) [17]. We also used
the dMRI measurements to perform probabilistic tractography and found the
major white matter bundles in every individual [20]. We focused on dMRI-based
tissue properties within the FLAIR-defined WMH and pursued a two-fold multi-
variate analysis strategy: (1) Using unsupervised learning to characterize WMH
differences across individuals and (2) focusing specifically on callosal white matter bundles that are particularly prone to WMH along their trajectory to better
understand the characteristics of periventricular WMHs.

1.1   Related Work

Previous studies that examined the relationship between dMRI measurements
and FLAIR [10,22,23] had some limitations that are addressed by the current
study. First, the data collected in these previous studies typically had lower
angular resolution and a single b-value shell (e.g., $b = 1000 \, \text{s/mm}^2$), limiting them
to rely on diffusion tensor imaging [4]. The data described in the present work was
collected with multiple directions and b-values ($b = 500, 1000, 2500 \, \text{s/mm}^2$), thus
making them amenable to advanced modeling techniques. In the few previous
cases where multiple b-values were used, those works focused specifically on MS
[25,30]. Second, previous work focused on broadly defined anatomical regions,
whereas the present work used recent developments to automatically delineate
precise WMH ROIs for analysis and methods for automatic delineation of major

white matter bundles [20]. Moreover, our two-pronged analysis allows both a general characterization of dMRI tissue property phenomenology within WMH, and a more detailed characterization of dMRI tissue properties along one bundle, as it relates to the FLAIR signal in the same bundle. The latter provides a specific biophysical interpretation of the WMH characteristics in this anatomical location.

2   Methods

2.1   Participants

48 participants (ages 70–103, mean age $= 79.81$; 26 females) participated in this study. The institutional review board at the Kaiser Permanente Research Institute approved the study protocol.

2.2   MRI Acquisitions

Data were acquired at the University of Washington at the Diagnostic Imaging Sciences Center (DISC) on a 3T Philips Ingenia Elition MRI scanner with a 32-channel head coil. A T1-weighted (T1w), T2-weighted (T2w), and FLAIR structural images and two diffusion-weighted acquisitions of opposite phase encoding directions were collected from each participant.

T1w. 3D MPRAGE T1w images were acquired at $1\,\text{mm}^3$ isotropic resolution (TR $= 6.5$ ms, TE $= 2.9$ ms, flip angle $= 9°$, FOV $= 256 \times 256$ mm, matrix size $= 256 \times 256$, 211 sagittal slices).

T2w. 3D spin-echo T2w images were acquired for each participants (TR $= 2500$ ms, TE $= 331$ ms, flip angle $= 90°$, FOV $= 256 \times 256$ mm, matrix size $= 288 \times 288$, voxel size $= 0.89 \times 0.89$ mm, slice thickness $= 1$ mm, 175 sagittal slices).

FLAIR. 3D FLAIR images were acquired with an inversion recovery sequence at $1\,\text{mm}^3$ isotropic (TR $= 5000$, TE $= 291$ ms, TI $= 1800$ ms, flip angle $= 90°$, FOV $= 256 \times 256$ mm, matrix size $= 256 \times 256$, 176 sagittal slices).

Diffusion-Weighted. Diffusion-weighted images were acquired with a spin-echo echo-planar imaging sequence at an in-plane spatial resolution of $1.8128 \times 1.8125\,\text{mm}^2$ (TR $= 3500$ ms, TE $= 89$ ms, FOV $= 232 \times 232$ mm, matrix size $= 128 \times 128$, slice thickness $= 2.20$ mm, slice gap $= 0.22$ mm, 57 axial slices). The dMRI data were collected at 3 b-values, 500, 1000, $2500\,\text{s/mm}^2$, with 32, 64, and 128 directions, respectively. Twenty-six non-diffusion-weighted ($b = 0$) measurements were interleaved with the diffusion-weighted measurements. Two diffusion-weighted images with opposite phase-encoding at all directions were acquired for each participant.

2.3   Preprocessing

T1w. T1w images were preprocessed using the QSIprep-0.18.1 [8] anatomical pipeline. In brief, the T1w image was corrected for intensity non-uniformity (INU) using `N4BiasFieldCorrection` (ANTs-2.4.3; [27]), and used as an anatomical reference throughout the workflow. The T1w image was reoriented into AC-PC alignment via a 6-DOF transform extracted from a full affine registration to the MNI152NLin2009cAsym template [12]. A nonlinear registration to the T1w image from AC-PC space was estimated via symmetric nonlinear registration (SyN) using `antsRegistration` (ANTs-2.4.3; [3]). Brain extraction was performed on the T1w image using SynthStrip [16] and automated segmentation was performed using SynthSeg [5,6]) from FreeSurfer-7.3.1.

T2w and FLAIR. First, T2w and FLAIR images were skull-stripped with `mri_synthstrip` [16] and corrected for intensity non-uniformity with the N4 algorithm [27] implemented in ANTsPy. Images were then coregistered to the participant's preprocessed T1w image from Sect. 2.3 through a rigid body registration. FLAIR images were standardized by participants' white matter voxel intensities based on the individual white matter masks created in Sect. 2.3. Lastly, nonlinear spatial normalization to the MNI152NLin2009cAsym template was performed on the T2w and FLAIR images.

Diffusion MRI. Diffusion MRI preprocessing was performed using the QSIprep-0.18.1 [8] diffusion pipeline. In summary, diffusion images were processed with MP-PGCA denoising (MRtrix3's `dwidenoise` [28], 5 voxel window) and Gibbs ringing correction (MRtrix3's `mrdegibbs` [19]). The mean intensity of the DWI series was adjusted so all the mean intensity of $b = 0$ images matched across each DWI scanning sequence, where any images with $b < 100$ s/mm^2 were treated as $b = 0$. Diffusion images were grouped by their phase-encoded polarity and merged into a single file. FSL-6.0.5.1 `eddy` was used for head motion correction and Eddy current correction (q-space smoothing factor $= 10$, 5 iterations, 1000 voxels; [2]). A linear first- and second-level models were used to characterize Eddy current related spatial distortion. q-space coordinates were forcefully assigned to shells and field offset was attempted to be separated from participant movement. Shell were aligned post-eddy correction. Eddy's outlier replacement was run by grouping data by slice, only including values from slices determined to contain at least 250 intracerebral voxels. Groups deviating by more than 4 standard deviations from the prediction had their data replaced with imputed values. Data was collected with reversed phase-encode blips, resulting in pairs of images with distortions going in opposite directions. Multiple DWI series were acquired with opposite phase encoding directions, so $b = 0$ images were extracted from each to be used for diffeomorphic registration. The susceptibility-induced off-resonance field was estimated from these pairs using a method similar to that described in [1]. The field maps were ultimately incorporated into the Eddy current and head motion correction interpolation. Final interpolation was performed using the `jac` method. DWI time series were resampled, generating a preprocessed DWI run with 1.8 mm isotropic voxels. Lastly, B1 field inhomogeneity correction was applied to the resampled images (MRtrix3's `dwibiascorrect` with N4 algorithm [27]).

2.4 Diffusion Modeling

Diffusional kurtosis imaging (DKI) [18] and free-water diffusion tensor imaging (FWDTI) [17] were fit using the Diffusion Imaging in Python (DIPY) software library [13–15]. We derived several metrics from each of these models. From DKI, we derived fractional anisotropy (DKI-FA), mean diffusivity (DKI-MD), and mean diffusional kurtosis (DKI-MK). In addition, the parameters of the DKI model were used to fit the White Matter Tract Integrity (WMTI) model [11], from which we derived a metric of axonal water fraction (DKI-AWF). From the FWDTI model, we derived FA (FWDTI-FA) and MD (FWDTI-MD), as well as the free water fraction (FWDTI-FWF).

2.5 Bundle Delineation

We used pyAFQ [20], an automated pipeline that identifies 24 major white matter bundles in individual brains, to perform bundle delineation and tractometry. The methods were previously described in detail in [20]; briefly: We used methods from DIPY [13] to perform constrained spherical deconvolution (CSD) [26] using all of the b-value shells in the dMRI acquisition, and we used the fiber orientation distribution functions in every voxel as cues for probabilistic tractography, with 8 streamlines seeded in every voxel in the white matter. Every individual's brain was aligned to the MNI152NLin2009cAsym template using non-linear registration, implemented in DIPY. For every major white matter pathway, inclusion and exclusion ROIs defined in the space of the template were back-transformed into the space of the individual and used to select streamlines that passed through inclusion and did not pass through exclusion ROIs. Each streamline was resampled to 100 points. Streamlines that deviated significantly (> 5 standard deviations) from the median trajectory were removed from each bundle. Tract profiles of tissue properties were derived by weighting the contribution of each streamline at each node inversely weighted by that node's distance from the median node in that position.

2.6 FLAIR White Matter Segmentation

HyperMapper [24] is a convolutional neural network-based segmentation algorithm that we used to generate probabilistic voxel-wise maps of WMH from the spatially normalized T1w, T2w, and FLAIR images. Voxels with probabilities ≥ 0.5 were defined as WMH voxels. Separate WMH ROIs were defined as contiguous voxels with scikit-image [29]. The WMH ROIs were transformed from the MNI152NLin2009cAsym template space to the space of the individuals and used in the remaining analyses.

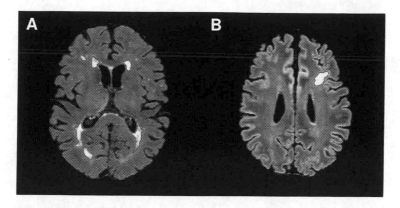

Fig. 1. Examples of FLAIR WMH segmentation. *(A)* Example of WMH segmentation in anterior and posterior periventricular regions. *(B)* Example of WMH segmentation in deep white matter.

3   Results

3.1   Automated WMH Segmentation

We found that automatic WMH segmentation produced from 5 to 87 (mean = 23, median =18) WMH ROIs. On average, the WMH voxels were 0.06% of the total white matter volume, with the most severe participant at 3.35% of their white matter considered hyperintense. WMH ROIs were most commonly located around anterior and posterior periventricular regions. Additionally, WMH ROIs were found in deep white matter (see Fig. 1).

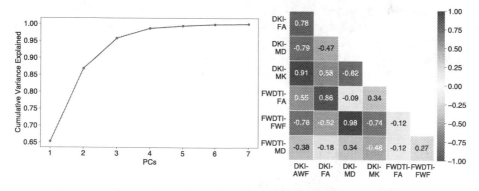

Fig. 2. Relationship between dMRI metrics within WMH ROIs. *(Left)* The first two PC axes cumulatively explained 86.84% of the variance. *(Right)* Correlation matrix of the dMRI metrics within WMH ROIs.

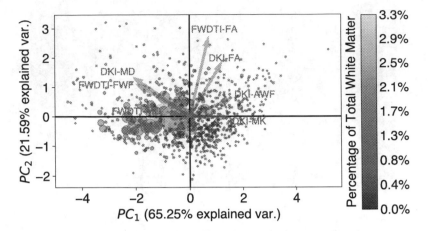

Fig. 3. PCA of dMRI metrics within WMH ROIs. Each dot represents a WMH ROI along the first two principal components. Dot size and color correspond to proportion of white matter the WMH ROI is of the participant's total white matter volume. The red vectors reflect the projection of dMRI metric axes into PC space. (Color figure online)

3.2 Principal Components Analysis (PCA)

We used PCA to decompose the diffusion metrics derived from the DKI and FWDTI models and examine how these metrics projected into linearly independent PC space. The first PC explained 65.25%, and the second PC explained 21.59% of the variance in the dMRI metrics within WMH ROIs. Cumulatively, the first two PCs explained 86.84% of the variance in the dataset (Fig. 2, *left*).

Figure 3 shows the projection of WMH ROIs and dMRI metric axes into PC space. The red arrows show the clustering of dMRI metrics, based on the correlations between the metrics shown in Fig. 2. While the WMH did not cluster into distinct clusters based on the PCA decomposition, we sought to interpret the distribution of WMH throughout this lower-dimensional projection, by categorizing WMH ROIs based on the quadrant into which they fell and took the average normalized dMRI metrics across ROIs in each quadrant.

Figure 4 shows the average normalized dMRI metrics within each PC quadrant, relative to the WM outside of WMH (z-scores, where 0 indicated values identical to normal-appearing WM). In $-PC_1/+PC_2$, there was higher DKI-MD, FWDTI-MD, and FWDTI-FWF, and lower DKI-MK and lower DKI-AWF. In $+PC_1/+PC_2$, there was higher DKI-FA and FWDTI-FA. In $-PC_1/-PC_2$, there was higher DKI-MD and FWDTI-MD along with lower DKI-FA, FWDTI-FA and DKI-MK, as well as lower DKI-AWF. In $+PC_1/-PC_2$, there was almost no difference from normal appearing WM, except for small reductions in DKI-AWF and FWDTI-FA.

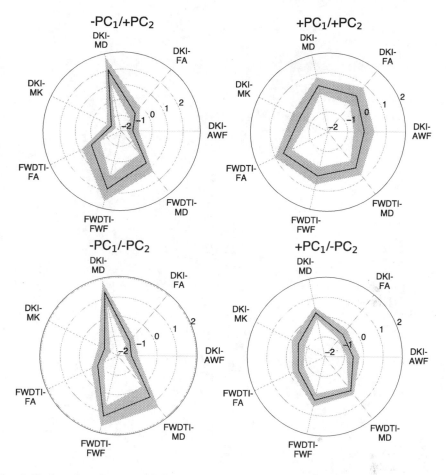

Fig. 4. Radar plot of average diffusion metrics by PC quadrant. Values near 0 represent no difference from participant's mean white matter values.

3.3   FLAIR and dMRI Metric Bundle Profiles

One of the types of WMH we identified in Sect. 3.2 ($-PC_1/+PC_2$) was reliably found around posterior periventricular areas and intersected mostly with the temporal callosal bundle. Therefore, when transitioning to the coordinate frame of the bundles, we focused on the temporal callosal bundle profile to characterize FLAIR WMH and its covariance with dMRI metrics across individuals.

We categorized participants into "high" and "low" FLAIR tracts by performing a median split based on the peak FLAIR value along the temporal callosal bundle (median z, relative to normal-appearing WM = 0.97; see Fig. 5). Corresponding dMRI metrics were also separated by the "high" vs. "low" peak FLAIR participants along the temporal callosal bundle.

As expected from the PCA analysis, we found that along the temporal callosal bundles of participants with high peak FLAIR values, lower DKI-MK, DKI-

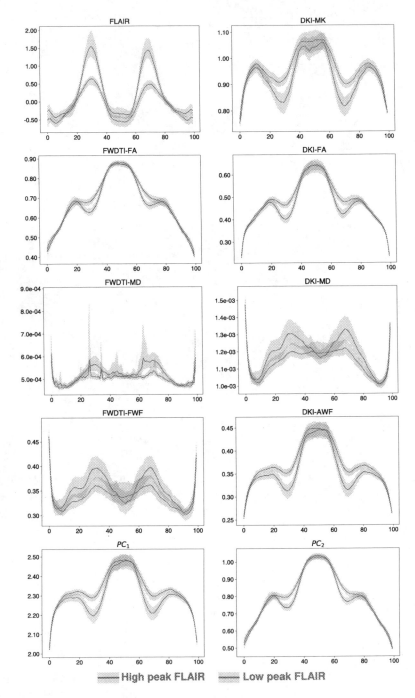

Fig. 5. FLAIR, dMRI metrics, and PC profiles along the temporal callosal bundle. Red and blue lines corresponded to participants with high and low peak FLAIR values along the temporal callosal bundle, respectively. (Color figure online)

FA, and DKI-AWF, as well as higher DKI-MD and FWDTI-FWF, as well as higher FWDTI-MD spatially corresponded to increases in relative FLAIR values The spatial pattern of the first two PC components along the temporal bundle were also remarkably similar to DKI-MK, FWDTI-FA, DKI-FA, and DKI-AWF profiles.

4   Conclusion

Using a large, high-quality dataset, focused on an aging population, we investigated the relationships between FLAIR WMH and dMRI metrics from advanced diffusion models.

First, using unsupervised learning, we found that > 86% of the variance in dMRI metrics within WMHs were explained by 2 principal components. Using these two components, we were able to classify the WMH regions of interest into four types: (1) Higher MD and FWF, (2) Higher FA, (3) Higher MD with lower FA, and (4) No difference from normal appearing WM. Previous work that used PCA to analyze white matter tissue properties in dMRI measurements demonstrated that much of the variance in these signals, even in healthy white matter is captured by two principal components [7].

To more closely link FLAIR measurements with a biophysical interpretation of the dMRI signal, we examined the tract profiles in the temporal callosal bundle, which passes posterior periventricular WMH. We found that higher FLAIR signal spatially corresponded to lower MK, FA, and AWF, as well as higher MD (measured both with the DKI and FWDTI models) and free-water fraction. Taken together, this pattern is consistent with the loss of axonal tissue and intrusion of CSF into the tissue from the ventricles in this WMH. We note that dMRI metrics may be less readily interpretable in WMH, because these parts of the tissue do not exhibit the same biophysical properties of normal-appearing white matter, which is where most studies using dMRI are conducted. In addition, these results are perhaps not surprising, but the quantification of these effects using these methods is novel and of potential utility. Specifically, future work with the ACT dataset will link the relationships that we have discovered here between FLAIR and dMRI metrics with: (1) stages of pathology (as quantified in FLAIR, e.g., using the Fazekas scale [9]), and (2) behavioral differences between individuals.

Acknowledgments. This research was funded by the National Institute on Aging (NIA; U19AG066567). Data collection for this work was additionally supported, in part, by prior funding from the NIA grants U01AG006781 and RF1AG056326. All statements in this report, including its findings and conclusions, are solely those of the authors and do not necessarily represent the views of the National Institute on Aging or the National Institutes of Health. We thank the participants of the Adult Changes in Thought (ACT) study for the data they have provided and the many ACT investigators and staff who steward that data. You can learn more about ACT at: https://actagingstudy.org/.

References

1. Andersson, J.L., Skare, S., Ashburner, J.: How to correct susceptibility distortions in spin-echo echo-planar images: application to diffusion tensor imaging. Neuroimage **20**, 870–888 (2003)
2. Andersson, J.L., Sotiropoulos, S.N.: An integrated approach to correction for off-resonance effects and subject movement in diffusion MR imaging. Neuroimage **125**, 1063–1078 (2016)
3. Avants, B., Epstein, C., Grossman, M., Gee, J.: Symmetric diffeomorphic image registration with cross-correlation: evaluating automated labeling of elderly and neurodegenerative brain. Med. Image Anal. **12**, 26–41 (2008)
4. Basser, P., Mattiello, J., Lebihan, D.: Estimation of the effective self-diffusion tensor from the NMR spin echo. J. Magn. Reson. **103**, 247–254 (1994)
5. Billot, B., et al.: SynthSeg: segmentation of brain MRI scans of any contrast and resolution without retraining. Med. Image Anal. **86**, 102789 (2023)
6. Billot, B., Magdamo, C., Cheng, Y., Arnold, S.E., Das, S., Iglesias, J.E.: Robust machine learning segmentation for large-scale analysis of heterogeneous clinical brain MRI datasets. Proc. Natl. Acad. Sci. **120** (2023)
7. Chamberland, M., et al.: Dimensionality reduction of diffusion MRI measures for improved tractometry of the human brain. Neuroimage **200**, 89–100 (2019)
8. Cieslak, M., et al.: QSIPrep: an integrative platform for preprocessing and reconstructing diffusion MRI data. Nat. Meth. **18**, 775–778 (2021)
9. Fazekas, F., et al.: CT and MRI rating of white matter lesions. Cerebrovasc. Dis. **13**, 31–36 (2002)
10. Ferris, J.K., et al.: In vivo myelin imaging and tissue microstructure in white matter hyperintensities and perilesional white matter. Brain Commun. **4**, fcac142 (2022)
11. Fieremans, E., Jensen, J.H., Helpern, J.A.: White matter characterization with diffusional kurtosis imaging. Neuroimage **58**, 177–188 (2011)
12. Fonov, V., Evans, A., McKinstry, R., Almli, C., Collins, D.: Unbiased nonlinear average age-appropriate brain templates from birth to adulthood. Neuroimage **47**, S102 (2009)
13. Garyfallidis, E., et al.: Dipy, a library for the analysis of diffusion MRI data. Front. Neuroinform. **8**, 8 (2014)
14. Henriques, R.N., et al.: Diffusional kurtosis imaging in the diffusion imaging in python project. Front. Hum. Neurosci. **15**, 675433 (2021)
15. Henriques, R.N., Rokem, A., Garyfallidis, E., St-Jean, S., Peterson, E.T., Correia, M.M.: [Re] Optimization of a free water elimination two-compartment model for diffusion tensor imaging. ReScience **3**, #2 (2017)
16. Hoopes, A., Mora, J.S., Dalca, A.V., Fischl, B., Hoffmann, M.: SynthStrip: skull-stripping for any brain image. Neuroimage **260**, 119474 (2022)
17. Hoy, A.R., Koay, C.G., Kecskemeti, S.R., Alexander, A.L.: Optimization of a free water elimination two-compartment model for diffusion tensor imaging. Neuroimage **103**, 323–333 (2014)
18. Jensen, J.H., Helpern, J.A., Ramani, A., Lu, H., Kaczynski, K.: Diffusional kurtosis imaging: the quantification of non-gaussian water diffusion by means of magnetic resonance imaging. Magn. Reson. Med. **53**, 1432–1440 (2005)
19. Kellner, E., Dhital, B., Kiselev, V.G., Reisert, M.: Gibbs-ringing artifact removal based on local subvoxel-shifts. Magn. Reson. Med. **76**, 1574–1581 (2016)
20. Kruper, J., et al.: Evaluating the reliability of human brain white matter tractometry. Aperture Neuro **1** (2021)

21. Kukull, W.A., et al.: Dementia and Alzheimer disease incidence: a prospective cohort study. Arch. Neurol. **59**, 1737–1746 (2002)
22. Min, Z.G., et al.: Diffusion tensor imaging revealed different pathological processes of white matter hyperintensities. BMC Neurol. **21**, 128 (2021)
23. Mito, R., et al.: In vivo microstructural heterogeneity of white matter lesions in healthy elderly and Alzheimer's disease participants using tissue compositional analysis of diffusion MRI data. NeuroImage Clin. **28**, 102479 (2020)
24. Mojiri Forooshani, P., et al.: Deep Bayesian networks for uncertainty estimation and adversarial resistance of white matter hyperintensity segmentation. Hum. Brain Map. **43**, 2089–2108 (2022)
25. Preziosa, P., et al.: NODDI, diffusion tensor microstructural abnormalities and atrophy of brain white matter and gray matter contribute to cognitive impairment in multiple sclerosis. J. Neurol. **270**, 810–823 (2023)
26. Tournier, J.D., Calamante, F., Connelly, A.: Robust determination of the fibre orientation distribution in diffusion MRI: non-negativity constrained super-resolved spherical deconvolution. Neuroimage **35**, 1459–1472 (2007)
27. Tustison, N.J.: N4ITK: improved N3 bias correction. IEEE Trans. Med. Imaging **29**, 1310–1320 (2010)
28. Veraart, J., Novikov, D.S., Christiaens, D., Ades-Aron, B., Sijbers, J., Fieremans, E.: Denoising of diffusion MRI using random matrix theory. Neuroimage **142**, 394–406 (2016)
29. van der Walt, S., et al.: The scikit-image contributors: scikit-image: image processing in Python. PeerJ **2**, e453 (2014)
30. Zhu, Q., et al.: The application of diffusion kurtosis imaging on the heterogeneous white matter in relapsing-remitting multiple sclerosis. Front. Neurosci. **16**, 849425 (2022)

Author Index